STUDIES IN THE FOUNDATIONS OF QUANTUM MECHANICS

STUDIES IN THE FOUNDATIONS OF QUANTUM MECHANICS

Edited by

PATRICK SUPPES
Stanford University

1980
PHILOSOPHY OF SCIENCE ASSOCIATION
EAST LANSING, MICHIGAN

Library of Congress Catalog Number 79-91232

Cloth Edition: ISBN 0-917586-12-3

Manufactured in the United States of America
by Edwards Brothers, Ann Arbor, Michigan

C O N T E N T S

PREFACE

The papers contained in the present volume are, mainly, written versions of reports given at the Stanford Seminar on the Foundations of Quantum Mechanics during the years 1975-1978. To a large extent, the authors of these papers are the same as the authors who appeared in the previous volume, _Logic and Probability in Quantum Mechanics_ (P. Suppes, Ed., Reidel, 1976). That volume covered the years 1972-1974 and was longer than this one; more of the papers originated outside the seminar. In the present case, all of the papers originated in the seminar with the exception of Demopoulos's paper, the commentary on it by Bub, and Wessels's paper, which was generated initially as a commentary on MacKinnon's paper.

From a formal academic standpoint, Nancy Cartwright and I share the responsibility of organizing the seminar, but it is fair to say that other regular participants take as much actual responsibility for its organization as do the two of us.

The twelve articles that appear in the present volume have been arranged in a semiconceptual way rather than in chronological order. The opening paper by MacKinnon contains a long historical analysis of the rise and fall of the Schrödinger interpretation of quantum mechanics. Wessels's, in turn, provides a detailed commentary on MacKinnon's main thesis about Schrödinger and offers an alternative analysis.

Noyes's article is concerned with an operational analysis of the double-slit experiment, which has been a subject of discussion in the seminar over several years. Given the intensity of our discussion of Noyes's paper and the continued controversies we have had over the double-slit experiment, future seminars should produce further reports on this matter. Noyes's paper offers a good sample of the issues that have concerned us.

Cartwright provides an interesting analysis of what it means to measure position probabilities and how the Born interpretation needs to be squared with the standard theorems of scattering theory. Her concern is to give a realistic theory of how measurement can actually take place and be consistent with the Born interpretation. The central point of her analysis is that measurement requires an exchange of energy between the object measured and the detector doing the measurement, and so we cannot simply speak abstractly of measuring position.

The most controversial paper in the volume is the one by Demopoulos on locality and the algebraic structure of quantum mechanics. Demopoulos attempts to show that locality, as defined by Bell and others, implies the algebraic homomorphism condition concerning the mapping of closed linear subspaces of a separable Hilbert space onto the two-element Boolean algebra, a matter which is itself closely connected with Gleason's well-known theorem that there are no two-valued measures on the partial Boolean algebra of the closed linear subspaces of a separable Hilbert space of three or more dimensions. Humphreys in his note on Demopoulos's paper shows that Demopoulos's claims are too strong when he asserts that there is no Boolean representation of a certain class of maximal magnitudes that arise in a natural way in Hilbert-space formalism. Bub in his short commentary shows that Demopoulos's argument that Bell's locality condition implies the existence of a homomorphism from the set of closed linear subspaces into a Boolean algebra, and thus into a two-element Boolean algebra, is incorrect. Demopoulos appends a Section V, entitled "Addenda," to his article to respond to these criticisms. Thus, after reading the comments by Humphreys and Bub, the reader should return to page 137, on which Section V of Demopoulos's paper begins.

Norman examines the consequences of assuming spontaneous projection postulates, which lead to a theory of spontaneous reduction of a system that is undisturbed by measurement. He shows that in a certain definite sense such a theory cannot have the same empirical content as standard quantum mechanics. Shiveley picks up on some of the earlier discussions in the seminar, especially a presentation given by Janet Beehner in 1977, to display an example of a nondistributive logic that avoids the undesirable properties of the usual special case.

Suppes and Zanotti continue their work reported in the earlier volume on a probabilistic analysis of hidden variable theories. In the present article they give a general argument requiring no detailed computations from quantum mechanics. They use de Finetti's principle of exchangeability and the principle of identity of conditional distributions under a hidden variable to show that there can be no hidden variable theory satisfying these two principles, together with negative correlation of observable variables such as spin, and locality in the sense of conditional statistical independence.

Domotor outlines a general program for developing a quantum decision theory. This is one of the few papers in the literature to combine statistical decision theory and the kind of probabilistic structures that arise in standard quantum mechanics.

The final article is really a bibliography on quantum logic, put together by Janet Beehner, who has given reports on quantum logic several times in the seminar. Her bibliography covers 381 items, consisting of either articles or books. It includes relevant materials from Donald Nilson's bibliography, published in the 1976 volume.

In preparing this volume for publication, I am very much indebted to Mrs. Dianne Kanerva, who not only was responsible for much of the editing and the preparation of the index, but also for the preparation of the camera-ready copy. I also thank Mrs. Marguerite Shaw for her assistance in various aspects of preparing the volume.

Patrick Suppes

THE RISE AND FALL OF THE SCHRÖDINGER INTERPRETATION

Edward MacKinnon
California State University, Hayward

I. THE ORIGIN OF WAVE MECHANICS

The formal development of Schrödinger's wave mechanics is familiar not only from recent historical accounts (Jammer, 1966, pp. 255-280; Scott, 1967; Gerber, 1969; Hund, 1974, pp. 147-153; Medicus, 1974, pp. 38-45), but even more from the fact that his methods and calculations have become an integral part of modern physics. Rather than give one more summary account of this formalism, we shall concentrate almost exclusively on the problems Schrödinger encountered in his attempts to give this formalism an adequate and consistent physical interpretation. Before doing this, however, we should situate Schrödinger's work with respect to the developments considered elsewhere.[1]

Schrödinger's wave mechanics was developed in a series of four papers, "Quantisierung als Eigenwertproblem" (1926b, 1926c, 1926d, 1926e; following Schrödinger's usage, these will be referred to as Q1, Q2, Q3, and Q4), and in two articles separate from this series but written during the same period.[2] Since neither these papers nor the available unpublished material explains how Schrödinger first formulated his wave equation, there have been different conjectures on this point. According to the most widely held view,[3] the fundamental equation of wave mechanics was suggested by Hamilton's analogy between ordinary mechanics and geometrical optics and by the further idea that, just as geometrical optics has to be replaced by a wave theory to treat fine-scale optical phenomena, so de Broglie's 'ray mechanics' had to be replaced by a quantum wave theory when treating phenomena of atomic dimensions. Though this interpretation stems from Q2 and from Schrödinger's later (1928) redevelopment of wave mechanics, it runs into serious difficulties when

Studies in the Foundations of Quantum Mechanics, pp. 1-57

considered as a chronological rather than a pedagogical account.

A different interpretation was suggested by Martin Klein (1964) and is being developed by Linda Wessels.[4] In essence, this is the contention that Schrödinger's initial ideas on wave mechanics stemmed from his work on a quantum theory of an ideal gas and that the adaptation and extension of the analogy Hamilton established between mechanics and optics was a later rationalization of ideas first developed in a more intuitive way. I find this a much more plausible interpretation but will defer to Wessels and others for detailed arguments on the comparative merits of the two interpretations.

My purpose here is not so much to explain where and how Schrödinger got his original interpretation, as to consider the changing use made of it in the development of his articles. Since, unfortunately, Schrödinger has left almost nothing in the way of an autobiographical account of his moves and their motives, it is necessary to have recourse to conjectures. The conjecture that supplies an interpretative framework for the account that follows can be stated fairly succinctly. Schrödinger's work on gas theory and his adaptation of de Broglie's concept of matter waves led him to the idea that particles could be represented as singularities in wave packets. After some searching, Schrödinger developed a wave equation to fit this assumption. When this yielded correct results for the hydrogen spectrum, Schrödinger knew that he was on the threshold of a major breakthrough.

In spite of this success, the interpretation of electrons as singularities in wave packets presented formidable difficulties which Schrödinger was not able to overcome. The strategy he followed in this impasse was quite similar to the strategy Heisenberg had employed some six months earlier when he ran into corresponding difficulties in developing quantum mechanics.[5] Schrödinger simply presented his new equation without attempting to justify it, worked out particular solutions, and showed how they could be interpreted in terms of observable phenomena. In QI and Q2 the basic formalism was developed in such a way that its intelligibility did not depend on an acceptance of the wave interpretation. Schrödinger certainly did not suppress this interpretation the way Heisenberg had suppressed his reliance on the virtual oscillator model. However, the problem of a model adequate to this new formalism was removed from a foundational role, relegated to the status of a side issue, and treated in an article that was not a part of the basic quantization series.

After doing this, Schrödinger proved to his own satisfaction--or, perhaps, to his own dissatisfaction--that his mathematical formulation was equivalent to Heisenberg's. This, I believe, precipitated something of a personal crisis. Unless Schrödinger's new theory could be shown to have a physical interpretation different from and superior to the interpretation of Heisenberg's formalism, then Heisenberg, rather than Schrödinger, would have to be accepted as the architect of the new age in quantum physics. Accordingly, in the third, and especially in the fourth, article of his quantization series, Schrödinger endeavored to develop and justify a physical interpretation of his mathematical formalism. This was a new interpretation that differed in significant respects from the interpretation of electrons as wave packets that Schrödinger had originally relied on. This new interpretation was unsatisfactory, and Schrödinger's commitment to it was ultimately self-defeating. Rather than accept the radical reinterpretation of his own results that Born, Bohr, and others developed, Schrödinger effectively opted out of the developments to which he had made such a prominent contribution. This interpretation, it must be admitted, is unprecedented and quite conjectural. Nevertheless, I believe that it does lend a plausible coherence to the account that follows.

It seems that Schrödinger first became acquainted with de Broglie's ideas through a suggestion of Pieter Debye,[6] then his colleague in Zurich, that he give a seminar on de Broglie's thesis. Einstein's enthusiastic endorsement was the decisive factor inducing both men to take de Broglie's ideas seriously. Schrödinger had been working on the quantum theory of ideal gases. After reading Einstein's papers, Schrödinger redeveloped Einstein's theory, introducing significant modifications in both the interpretation of an ideal gas and the interpretation of the particles that compose it. This is the only aspect of Schrödinger's gas theory that we will consider.

Though influenced by Einstein and de Broglie, Schrödinger did not accept either of their versions of the light-quantum hypothesis or gas theory. It is more natural, Schrödinger (1926a) argued, to apply the new statistics to the degrees of freedom of the black-box cavity itself, or the ether resonations, than to light particles. Similarly, instead of assigning energies to molecules in a gas, one can quantize modes of vibration, assigning to the sth mode the possible energies, $0, \varepsilon_s, 2\varepsilon_s, 3\varepsilon_s, \ldots, n\varepsilon_s$, depending on whether it is occupied by 0, 1, 2, ..., n molecules. Though this is similar to de Broglie's idea of many molecules sharing the same phase wave, there are, nevertheless, some significant differences. The de Broglie waves move at a velocity of v^2/c relative to the molecular velocity v; the Schrödinger waves are stationary. The de Broglie waves are coupled with particles in a dualistic system; the Schrödinger waves are introduced to replace particles. At this stage, however, Schrödinger presented the replacement as a suggestion rather than as a developed doctrine, a way of applying the new statistics to the bearers of the energy states rather than to the energy states themselves.

Schrödinger was treating an enclosed gas as a unified whole, almost as if it were an organic unit. This, in turn, necessitated a reinterpretation of the molecules that make up a gas. Schrödinger's reinterpretation hinged on the way he adapted de Broglie's ideas. A molecule of rest mass m and velocity $v = c\beta$ is, as he interpreted de Broglie, nothing but a signal, the crest of the froth (Schaumkamm) of a wave system whose frequency ν lies in the neighborhood of $\nu = mc^2/h\sqrt{1 - \beta^2}$ and whose phase velocity u is governed by the dispersion law $u = c/\beta = c^2/v$. These are not the de Broglie relativistic phase waves but the waves that make up a group in a dispersive medium. Even the relativistic formulation itself simply serves as a bridge between de Broglie's work and his own. Once introduced, it is quickly replaced by a different formulation, one in which molecules are treated nonrelativistically and radiation relativistically (but not the way de Broglie treated it). With this, Schrödinger is able to reproduce Einstein's ideal gas results, interpreting them in terms of vibration states of the gas as a whole rather than as the statistical interaction of individual molecules. Though the mathematics is equivalent to Einstein's, the interpretation of molecules in terms of waves in a dispersive medium presents a problem to which

Schrödinger returned in the concluding section of his article.

The question considered there is whether it is possible to treat molecules or light quanta in terms of the interference of plane waves. A superposition of a great number of such waves with a common wave normal and with near-neighboring frequencies can serve as a signal that is sharply localized. But, Schrödinger wondered, will it stay localized, contracted in a small space, especially in the case of three-dimensional motion? This is the problem the paper raised but did not resolve. Schrödinger cited some earlier work of Debye and von Laue indicating that an affirmative answer is possible only when the frequency difference is infinitesimal. In other cases, the singularity passes through a focal point and then disperses. Schrödinger's (1926a) conclusion is worth quoting: "If one can avoid this consequence through a quantum theoretical modification of the classical wave law then it appears that a path is prepared for the solution of the light-quantum dilemma" (p. 101).

This paper was finished some six weeks before the first paper in Schrödinger's series on wave mechanics. It clearly shows that Schrödinger was favorably disposed toward a physical interpretation of particles as wave packets but acutely aware of the dispersion difficulties this presented. It undoubtedly seemed prudent to minimize the physical interpretation and rely on the development of an adequate mathematical formalism. Yet this, too, presented formidable difficulties. Schrödinger later told Dirac, with whom he shared the Nobel Prize, that he spent months trying to develop a relativistic wave equation but despaired when the equations he developed would not yield the correct results for the hydrogen atom.[7] Then he switched to a nonrelativistic equation, probably intending it as an approximation to the relativistic formulation that the situation seemed to require. This approximate equation quickly proved an unprecedented success in supplying a unified solution for problems that could only be solved piecemeal, if at all, by older formulations.

This paradoxical situation explains, I believe, the peculiar way in which Q1 and Q2 are developed. Schrödinger had an equation that worked but could not give any satisfactory reason, apart from this success, why this equation should obtain at all. It could not be developed from general principles, though some principles from classical physics eventually supplied a suggestive point of departure. Nor was it in accord with the principles of relativity that should apply to an exact solution. On the other hand, the equation could not be given the type of semi-intuitive physical justification that physicists regularly employ in setting up differential equations to cover physical situations. This would involve reliance on a novel interpretation of particles that even Schrödinger, then the sole supporter of this view, recognized as unproved and perhaps inconsistent. The best way out seemed to be to present the equation and postpone its justification. This is essentially what Schrödinger did in Q1.

II. THE DEVELOPMENT OF WAVE MECHANICS

Schrödinger's first quantization paper (Q1) attempted to show, for the simple case of the nonrelativistic and unperturbed hydrogen atom, that the usual quantization procedure can be replaced by another postulate in which integers are not assumed but follow in a natural way, as, for example, in the number of nodes of a vibrating string. This hint of a wave analogy, together with a brief allusion to vibrations at the conclusion of the article, supplies the only textual basis for interpreting Schrödinger's new equation as a wave equation. The emphasis in the first two communications is on the fact that the new method of quantizing yields integers in a natural way.

The usual form of the quantum conditions is connected with the Hamilton-Jacobi partial differential equation,

$$H(q, \partial S/\partial q) = E \, . \tag{1}$$

A solution for (1) is sought that is a <u>sum</u> of functions, each of a single one of the independent variables, q. Schrödinger modified this by introducing a new unknown ψ through the substitution

$$S = K \log \psi \, . \tag{2}$$

Here K is a constant that, like S, has the dimensions of action (or of h). The basic advantage resulting from this substitution is that ψ will now appear as a product of the functions of the individual coordinates rather than as a sum. With this substitution, (2) takes the form

$$H(q, K/\psi, \partial\psi/\partial q) = E \ . \tag{3}$$

Before developing any particular solutions for (3), Q1 stipulates the conditions that ψ must fulfill (Schrödinger, 1926b).[8] It must be real over the whole of configuration space, unique valued, finite, continuous, and twice differentiable.

These conditions of adequacy are simply presented with no justification, yet they deserve some comment. The most notable feature about them is that they are conditions imposed on ψ considered as a mathematical function. The physical interpretation to be accorded ψ is only indirectly operative through the stipulation that ψ must be real. The reason for this stipulation is undoubtedly the reason given in later parts of the series: If the ψ-function corresponds to something physically real, such as a matter wave, then it should be real rather than imaginary or complex. The force of this reason is somewhat blunted, however, by the fact that the ψ-function is in configuration space (a product space with three coordinates for each particle) rather than ordinary space. Though this is listed as a basic criterion for the acceptability of the ψ-function, it is, in fact, never fulfilled in this or any other article in the quantization series.

The rest of Q1 has a rather chiaroscuro quality. The solution of the Schrödinger equation, to use the now standard term, is developed with such elegance and generality that this treatment has become a stable part of physics ever since. Yet the justification of this virtuoso performance is so obscure that Schrödinger (1927a, p. 13), in the introduction to his next article, referred to it as an unintelligible transformation for an incomprehensible transition; Q2 is primarily concerned with making this transition more intelligible. This peculiar order of development would seem to reflect the fact that Schrödinger had more confidence in the conclusions flowing from his equation than in any justification he was then able to give for its introduction.

After the general specification of requirements that

ψ must fulfill, Q1 is chiefly concerned with solving equation (3) for the hydrogen atom. The method of solution employed need only be outlined here. By applying the variational method to (3), Schrödinger developed two equations. The second of these, concerning the behavior of an integral of ψ, $\partial\psi/\partial_{normal}$, is handled by requiring that physically significant quantities vanish in a suitable way at infinite distances. Though this might seem to be suggested by the wave interpretation of ψ, Schrödinger (1927a, pp. 7-8) relies on mathematical rather than physical reasons.[9] The first, and more familiar, equation is

$$\nabla^2\psi + 2m/K^2(E + e^2/r)\psi = 0 \ , \tag{4}$$

where K must, for numerical agreement, have the value h/2.

Solving equation (4) was far from routine. Schrödinger's was the first solution of a partial differential equation exhibiting both a continuous and a discrete eigenvalue spectrum.[10] One significant point about this solution was noted by Schrödinger himself (1927a, p. 9); the solution was developed in a way that was neutral with respect to interpretations of atomic structure. Though he would have preferred to relate ψ to some vibratory process within the atom, something analogous to beats in music, he relied instead on the fact that the correct numerical relations came out in a natural way without the imposition of arbitrary conditions.

Schrödinger's second quantization paper (1926c) related more directly to the work of de Broglie. The paper began with a clarification of the basis of Q1, showing how Schrödinger's variation principle corresponds to Fermat's principle for wave propagation, though Schrödinger's principle is in configuration space. Similarly, the Hamilton-Jacobi equation Schrödinger used in Q1 could be interpreted as expressing Huyghens' principle for wave propagation. The justification that Q2 presented can be described as a nonrelativistic reinterpretation of de Broglie's results. In place of de Broglie's relativistic phase waves, however, Schrödinger (1927a, p. 16) arrived at a system of wave surfaces that form a progressive but stationary wave motion in configuration space.[11]

The problems configuration space presents will be considered later. Of more immediate interest is the comparison drawn in Q2 (Schrödinger, 1927a, 25ff.): De Broglie's theory (rightly reinterpreted) stands to Schrödinger's theory as geometric optics stands to wave optics. Geometric optics represents a valid approximation when one is studying phenomena, such as reflection and refraction, for which wave effects are so negligibly small that light may be treated as a set of rays. Similarly, de Broglie's geometric wave mechanics (or ray mechanics) is an approximation no longer valid when one is treating phenomena of a size comparable to the wave lengths involved. In using such a reinterpretation of de Broglie's work, Schrödinger faced a basic problem of showing how his mathematical formulation of wave surfaces in configuration space related to de Broglie's theory of matter waves.

What Schrödinger took from de Broglie was actually the formulation based on wave velocity and group velocity, rather than the relativistic phase-wave formulation. Though the way Schrödinger handled this is quite involved mathematically, the basic method can be summarized in a fairly qualitative way. One postulates a group of waves and then sets up and solves the Hamilton-Jacobi equation to get a point of agreeing phase for a whole aggregate of wave groups. This singularity in configuration space takes the place of de Broglie's particle riding a wave, and its motion defines the geometric locus of points of agreeing phase. Schrödinger was unable to prove mathematically that a superposition of such wave disturbances really produces a notable disturbance only in a very small region--or that there is no spreading of the singularity. Accordingly, he postulated a lack of spreading and presented this as a physical, rather than a mathematical, hypothesis subject to the test of experimental trial (Schrödinger, 1927a, p. 25). The conclusion Schrödinger drew from this protracted comparison was that in dealing with small-range phenomena, where the ray-mechanics approximation is invalid, one must abandon any images of electrons moving along definite paths and rely on a wave equation. To back up this deliberate disregard of visualizable models, Schrödinger cited the work of Heisenberg, Born, and Jordan and noted that their strivings seemed to manifest the same tendency. Though he had not yet found the link connecting his development with theirs, Schrödinger (1927a, p. 30) expressed the hope that his own method would eventually lead to a more intuitive understanding of the microscopic mechanical processes.

After this rather abstract discussion, Schrödinger set up and solved his wave equation for some basic and now familiar problems: the Planck-Hertz oscillator, the rigid rotator with fixed axis and with free axis, and the nonrigid rotator. In solving these problems, he concentrated on achieving mathematical results while sidestepping the problem of the precise physical interpretation to be accorded these results. As he put it: "The question of how the energy is really distributed among the proper vibrations, which has not been taken into account here up till now, will, of course, have to be faced sometime" (p. 34).

III. THE INTERPRETATION OF WAVE MECHANICS

Schrödinger intended his four quantization papers as a series covering the whole foundation of wave mechanics, yet he interrupted this series to write two further papers--a short one concerned with the physical interpretation of the new wave mechanics and a much longer one on the relation of his new wave mechanics to Heisenberg's matrix mechanics. In the short paper (Schrödinger, 1926f), more a note than a regular paper, he attempted to show that, at least in one case, a group of proper vibrations behaves like a particle. The case chosen is a rather artificial one, a one-dimensional simple harmonic oscillator with frequencies confined to a specially selected and rather narrow range. The result he developed can be described as the motion of a tall, narrow Gaussian hump, corrugated with vibrations, which moves like a classical particle in that the wave packet does not spread out in the course of time.

Here, I believe, the attitude displayed is of more significance than the example, which never admitted of a significant generalization to the more realistic three-dimensional case. Schrödinger clearly hoped that a particle could be represented by a packet of waves, each of which satisfied his equation, and he kept trying to develop a mathematical expression for a physically realistic wave packet that would remain particle-like in that it would not spread out in time. Nevertheless, he kept this effort separate from his more basic quantization series. The mathematical formalism developed in that series did not require any doubtful physical interpretation as a foundation. All that the formalism required was one general partial differential equation, which could be adapted to particular systems by a suitable choice of variables and boundary conditions. This replaced both the equations of motion and the quantum conditions of the older quantum theory. Moreover, this equation yielded integers in a natural way, as eigenvalue solutions to a wave equation rather than through the postulation of discontinuities in atoms. Schrödinger had introduced the requisite equation and worked out solutions for basic cases. The Q3 and Q4 papers were projected extensions of this basic breakthrough to cover perturbation theory. But then Schrödinger made a new discovery which, I believe, induced him to alter the emphasis of the last two quantization papers, making the physical interpretation of the mathematical formalism a basic rather than a peripheral issue.

In Q2, Schrödinger had expressed the belief that the work of Heisenberg, Born, and Jordan manifested the same tendency as his own work. After this paper was completed, he considered the relation between the two theories in more detail and demonstrated the mathematical equivalence of the two systems. This demonstration presents both historical and philosophical problems. The philosophical problems will be considered later. The historical problem is the question of who first discovered the basic relation between Schrödinger's differential operators and Heisenberg's matrices.[12] There were at least two, and perhaps three or even four, roughly simultaneous discoveries of the same basic relation. To see the basis for this interrelation we should briefly consider some earlier work of Heisenberg and his correspondence with Wolfgang Pauli.

According to van der Waerden (1967, pp. 52-53), in September 1925, Heisenberg, Pauli, Dirac, and Hermann Weyl (Schrödinger's colleague in Zurich) independently developed the basic commutation relation

$$(5)\quad p_r q_s - q_s p_r = \delta_{sr} h/2\pi i, \text{ where } \delta_{sr} = \begin{cases} 1, s = r \\ 0, s \neq r \end{cases}.$$

On September 24, Heisenberg, who was then in Copenhagen, sent Pauli a postcard containing a brief proof that this set of commutation relations implies that

$$(6) \qquad 2\pi i/h(Fp_r - p_rF) = -\partial F/\partial q_s$$

for every function F that can be developed in a power series in q and p. This was taken over into the three-man paper on matrix mechanics in the form

$$(7a) \qquad fq - qf = (\partial f/\partial p)(h/2\pi i) \,,$$

$$(7b) \qquad pf - fp = (\partial f/\partial q)(h/2\pi i) \,.$$

These equations related matrices representing momentum p (or position q) to a differential operator for the conjugate position (or momentum) variable. When Schrödinger developed the relation between wave mechanics and matrix mechanics, he was familiar with the three-man paper and used the idea behind equation (7) as his point of departure.[13]

Independently, Pauli developed the same basic relationship and explained it in a letter sent to Pascual Jordan on April 12, 1926, prior to the publication of Schrödinger's article on May 4, 1926, a letter that has been made public only recently.[14] Though never published, this letter appears to have received some circulation. Thus, Gregor Wentzel (1926), in an article received for publication on June 18, 1926,[15] cited Pauli as one who discovered the mathematical equivalence of the two formalisms.

Similarly, Max Born, in an interview with Thomas Kuhn,[16] declared that he was amazed to see that Schrödinger could put quantum theory in the form of a differential equation. When he wrote to Pauli and asked what the relationship was between the two formalisms, Pauli replied that a Heisenberg matrix is equivalent to a Schrödinger integral $\int \psi_n f \psi_m \, dv$. Though the date of this letter was not given, it seems reasonable to assume that it was written before Schrödinger's paper was published. Schrödinger's paper was received on March 16, 1926, and published some two months later. Heisenberg, who was in constant communication with Pauli, was probably familiar with this relation prior to the publication of Schrödinger's paper. On May 26, 1926, he wrote a letter to

P. A. M. Dirac explaining that the Schrödinger orthonormal wave functions $\int \psi_n \psi_m \, d\Omega = \delta_{nm}$ are related to his matrix f, representing an observable, by the formula $f_{nm} = \int f \psi_n \psi_m \, d\Omega$, where $d\Omega$ is a volume element.[17] The source of this formula is not indicated. It might have been communicated by Pauli--a study of Pauli's correspondence might settle this--or it might have been developed independently. Carl Eckart's (1926) independent development of the same basic relation indicates that it was in the air.[18] Even if he had wished to do so, Schrödinger could not have long avoided a detailed comparison between wave and matrix mechanics.

Our primary consideration, however, is not with the question of historical priority but with the significance that the establishment of this relation between the two theories had on the development of Schrödinger's position. To bring this out, we will separate three topics that are mixed together in Schrödinger's original development: the formal correspondence between Schrödinger's wave functions and Heisenberg's matrices, the manner in which this allowed Schrödinger to extend the theory he was developing, and the peculiar problems of interpretation this comparison engendered.

The formal relationship between the two different approaches is so familiar to physicists from subsequent textbook treatments that it requires little elaboration. What is worth noting is that Schrödinger's treatment, though undoubtedly written in haste, is characterized by his usual balance of generality in scope and precision in detail. By introducing an arbitrary orthonormal set of functions $u_i(x)\sqrt{\rho(x)}$, where the density function $\rho(x)$ is defined by

(8) $$\int \rho(x) u_i(x) u_k(x) = \delta_{ik}$$

and x is an abbreviation for the product of spatial coordinates $q_1, q_2, \ldots, q_n$, one may set up the matrices for any two general functions

(9a) $$F^{k\ell} = \int \rho(x) u_k(x) [F, u_\ell(x)] \, dx ,$$

(9b) $$G^{\ell m} = \int \rho(x) u_\ell(x) [G, u_m(x)] \, dx .$$

Here $[F, u_\ell(x)]$ is not a Poisson bracket but a very general expression that for most functions reduces to $Fu_\ell(x)$ multiplied by $(-i\hbar)$ raised to the appropriate power. Schrödinger was able to show that the matrices so defined conform to the standard laws of matrix multiplication

(10a) $$(FG)^{kn} = \sum_\ell F^{k\ell} G^{\ell n} ,$$

that they fit the Hamiltonian equations proper to matrix mechanics

(10b) $$\begin{aligned} (dq_\ell/dt)^{ik} &= (\partial H/\partial p_\ell)^{ik} , \\ (dp_\ell/dt)^{ik} &= (-\partial H/\partial q_\ell)^{ik} , \end{aligned}$$

and that for the orthonormal functions in (8) one can use functions that are solutions of the equation (in modern notation), $H\psi = E\psi$. Less technically, it is possible, Schrödinger demonstrated, to begin with the Schrödinger equation and then define matrices by means of an integral of an operator and matched Schrödinger eigenfunctions. The resulting matrices obey all of Heisenberg's laws and yield his results.

Norwood Hanson (1963, pp. 113-134) has repeatedly, and I believe correctly, insisted that Schrödinger's proof did not establish the equivalence of the two theories but merely the intertranslatability of the two formalisms.[19] A physical interpretation is an essential part of a theory and, in this case, the interpretations accorded the two formalisms were quite different. However, neither Schrödinger nor Heisenberg had an adequate physical interpretation for the formalisms they had developed. Intertranslatability of formulas, accordingly, was the crucial issue. Schrödinger (1927a) expressed this rather emphatically:

> If the two theories--I might reasonably have used the singular--should be tenable in the form just given, i.e., for more complicated systems as well, then every discussion of the superiority of the one over the other has only an illusory object, in a certain sense. For they are completely equivalent from the mathematical point of view, and it can only be a question of the subordinate point of convenience of calculation. (p. 57)

The comparison of the two theories allowed Schrödinger to extend his previous results in two significant ways. First, to make the two theories agree in form, he had to introduce the now familiar operator substitutions $p_\ell \rightarrow -i\hbar\partial/\partial q_\ell$, a technique that proved extremely helpful in later extensions of wave mechanics. Second, since he could now include the work of Heisenberg, Born, and Jordan in his own theory, he automatically had a method of handling transition probabilities and line intensities.

The problems of interpretation this comparison engendered are not settled by the available textual material. To fill this lacuna, I am offering a conjecture on Schrödinger's motivation. A bit of background may bring out the significance of this conjecture. Since de Broglie and Schrödinger soon associated themselves with Einstein, Planck, and the older generation of physicists in opposing the statistical interpretation of quantum theory, it has become customary to think of them as classical physicists at heart, desiring theories that give realistic representations of objectively existing bodies. This is certainly the impression they gave after 1927. But to understand their behavior at this crucial turning point it is more helpful to consider the attitudes they displayed prior to 1926.

Few physicists seemed as willing to jettison established physical concepts and laws as de Broglie. He not only introduced matter waves, light-quanta with mass, light molecules made up of indefinitely large numbers of light atoms, and a radically new conception of gases; he was also prepared to jettison Maxwell's electromagnetic laws and other parts of classical physics that did not fit his new theory. Schrödinger, always a loner, guarded his reserve to the end. One can only speculate about the motives that drove him on during this, the most creative period of his long career. However, the grounds for speculation seem well enough established.[20] Schrödinger was strongly influenced by Boltzmann's students, Fritz Hasenörl and Franz Exner, and repeatedly declared that the statistical heritage of Boltzmann was his first love in physics. In the inaugural address (Schrödinger, 1957) he gave at the University of Zurich in 1922, "What is a Law of Nature?", he defended the idea that[21]

> it is quite possible that Nature's laws are of a thoroughly statistical character. The demand for an absolute law in the background of the statistical

> law--a demand which at the present day almost everybody considers imperative--goes beyond the reach of experience. Such a dual foundation for the orderly course of events in Nature is in itself improbable. The burden of proof falls on those who champion absolute causality, and not on those who question it. For a doubtful attitude in this respect is today the more natural. . . . I prefer to believe that, once we have discarded our rooted predilection for absolute Causality, we shall succeed in overcoming these difficulties, rather than expect atomic theory to substantiate the dogma of Causality. (p. 147; italics in original)

When Bohr, Kramers, and Slater (1924) developed the view that the conservation of energy is merely a statistical law on the atomic level, Schrödinger (1924) came out as one of their strongest supporters.[22] In autumn 1925, just before he began the development of quantum mechanics, Schrödinger wrote a short essay, "Seek for the Road," presenting a Vedantic view on the unity of consciousness. To counter the objection that Vedantic doctrine was communicated by allegories rather than by strict reasoning, Schrödinger (1964) cited the scientist's regular reliance on allegorical pictures:

> We intellectuals of today are not accustomed to admit a pictorial analogy as a philosophical insight; we insist on logical deduction. But, as against this, it may perhaps be possible for logical thinking to disclose at least this much: that to grasp the basis of phenomena through logical thought may in all probability be impossible, since logical thought is itself a part of phenomena, and wholly involved in them; and we may ask ourselves whether, in that case, we are obliged to deny ourselves that use of an allegorical picture of the situation, merely on the grounds that its fitness cannot be strictly proved. (p. 19)

Further on in the same essay he counters the objection that the organic is radically different from the inorganic by relating the distinction to the observer's perspective: "But despite such offences against current usage, it seems to me that the idea that the fundamental contrast between organic and inorganic lies not in the constitution of the object but in the attitude of the observer is well worth considering" (p. 43). Later, Bohr repeatedly used a very similar explanation of the relation between the organic and the inorganic to illustrate his doctrine of complementarity.

Schrödinger opposed, rather than fathered, the probabilistic interpretation of wave mechanics, however. For an initial understanding of his motivation, we shall ignore the more or less philosophical reasons later developed to defend a position he had already accepted and consider the conflict for priority. James Watson (1968), one of the codiscoverers of the structure of DNA, has vividly depicted the emotional intensity generated by the desire to be first with a fundamental discovery and the feverish activity this can inspire. Schrödinger realized that the new developments in quantum theory constituted not merely another breakthrough but a decisive turning point in modern physics. The master architect of this intellectual revolution would be the Newton of the new era. Schrödinger appears to have striven for this post with an intensity that was almost manic. Thirty-six years later, Born could still remember how furious Schrödinger was when Born first proposed a probabilistic reinterpretation of Schrödinger's ψ-function (see Kuhn et al., 1967, pp. 23-29). When Bohr decisively refuted his interpretation, a topic we will consider later, Schrödinger had such a physical collapse that he had to be confined to bed in Bohr's house. It is not too fanciful to interpret this as a depressive stage following a manic high. Schrödinger never again matched the creative brilliance of his technical productions in 1926--something with few parallels in the history of science.

Schrödinger was not the first with wave mechanics. But he had so changed and extended de Broglie's ideas that the final results were much more Schrödinger's than de Broglie's. Then Schrödinger proved that the wave mechanics formalism, his distinctive contribution to the new scientific revolution, was equivalent to the formalism that Heisenberg had already developed. Heisenberg, not Schrödinger, was the intellectual architect of the new era. This must have been a bitter blow. There was really only one way to avoid the deflating conclusion that Heisenberg, rather than Schrödinger, was the master builder of the new quantum theory. That was to draw a sharp distinction between the formal mathematical theory and its physical interpretation. If Schrödinger's theory led to a different and more adequate physical interpretation, then it should be considered more basic. Both Schrödinger and Heisenberg seem to have realized this immediately. Before developing a detailed comparison of his work with Heisenberg's, Schrödinger (1927a) summarized his initial reaction:

> I did not at all suspect any relation to Heisenberg's theory at the beginning. I naturally knew about his theory, but was discouraged if not repelled (abgestossen) by what appeared to me as very difficult methods of transcendental algebra, and by the want of perspectivity (Anschaulichkeit). (p. 46, note 1)

This criticism is a bit peculiar if interpreted as a straight scientific criticism. Schrödinger's paper on Heisenberg, written while he was developing his own theory, clearly shows that Schrödinger had no difficulty in mastering and reinterpreting Heisenberg's "very difficult methods." In sidestepping any direct reliance on a picture of particles as wave packets and relying on a transformed version of the Hamilton-Jacobi equation, Schrödinger was matching Heisenberg's lack of Anschaulichkeit.

Schrödinger's reaction, however, is quite intelligible when understood as part of a competition for priority. It is interesting to compare this with Heisenberg's reaction to Schrödinger's theory. In the summer of 1926, Heisenberg (1960) wrote a letter in which he said, "The more I ponder the physical part of Schrödinger's theory the more disgusting (abscheulicher) I find it" (p. 44). The physical part of Schrödinger's theory was the idea of matter as vibrations. Though Heisenberg now found this disgusting, he had used a somewhat similar idea, the virtual oscillator model, as the original basis for his own development of quantum mechanics (cf. MacKinnon, 1977).

Schrödinger (1927a) began his comparison article by indicating the areas of agreement between the two approaches:

> In what follows the very intimate inner connection between Heisenberg's quantum mechanics and my wave mechanics will be disclosed. From the formal

> mathematical standpoint, one might speak of the identity of the two theories. (p. 46; italics in original)

Though the formal mathematical parts of the two theories are equivalent, wave mechanics, Schrödinger argued, supports a different and clearer physical interpretation. However, his treatment of this difference in the comparison article is more a promissory note than a developed argument. A basic point made is that Schrödinger's theory accords a clear significance to the amplitude of the atom's electrical oscillations. This, in turn, makes the polarization and intensity of the emitted light intelligible on the basis of the Maxwell-Lorentz theory. This manner of interpreting the mathematical formalism implied a reliance on unobservable elements, something Heisenberg had officially repudiated with his stress on the primacy of observables in scientific theories. Schrödinger, accordingly, rejected any theory of scientific explanation that held that the task of a scientific theory is essentially the establishment of an economical mathematical description of the empirical connections between observed quantities. To buttress the role he assigned unobservables, Schrödinger (1927a, p. 53) cited examples of physical problems that can hardly be treated without a reliance on intuition: how two colliding atoms rebound from one another and how an electron or an α-particle is diverted when it is shot through an atom with a given velocity and with an initial path at a given perpendicular distance from the nucleus.

These examples, unfortunately, do little to further Schrödinger's argument. The intuitive view physicists bring to such atomic collisions, or near misses, is one of electrons and atoms as particles rather than waves. Even this intuition may be effectively discarded when the equations have been set up. To make his case, Schrödinger clearly needed something more than a general argument on the role of intuition in physical reasoning. He needed a developed physical interpretation that differed from any physical interpretation supported by Heisenberg's matrix mechanics. This he thought he might have in the coupling of the atom to the electromagnetic field. In a matrix approach, even the electromagnetic field should be represented by matrices. In wave mechanics one could couple ψ, a scalar, to the electromagnetic field represented by the familiar Maxwell-Lorentz equations. This problem took on an increasing importance in Schrödinger's writings. If he could establish the sort of solution he desired, he would also establish the superiority of wave mechanics. Accordingly, we must consider the details of his treatment and the special difficulties they presented.

His basic conjecture was that the spatial distribution within the atom is given by the real part of $\psi\partial\bar{\psi}/\partial t$. When ψ is expanded in the form $\psi = \Sigma_k C_k U_k(x)\exp(2\pi i/E_k t)$, then the real part of $\psi\partial\bar{\psi}/\partial t$ is

(11)

$$R_e(\psi\partial\bar{\psi}/\partial t)$$

$$= 2\pi \sum_{k,m} C_k C_m((E_k - E_m)/h)u_k(x)u_m(x)\sin 2\pi(t/h)(E_m - E_k) .$$

Since the frequencies are low, the radiation can be estimated from the dipole moment. To get this, Schrödinger multiplied (11) by a Cartesian coordinate q_ℓ and a density function $r^2\sin\theta$ and integrated. This yields for the component of the dipole moment in the direction q_ℓ

(12)

$$Mq_\ell = 2\pi \sum_{k,m} c_k c_m q_\ell^{km}((E_k - E_m)/h)\sin 2\pi(t/h)(E_m - E_k) ,$$

where q_ℓ^{km} is the Schrödinger equivalent of the Heisenberg matrix element $q_\ell^{km} = \int u_k q_\ell u_m\, dx$. Equation (12) gives, as spectroscopy requires, a Fourier expansion of the atom's electrical moment in which only term differences appear as frequencies. What is more important for Schrödinger's purposes, it seemed to make some of Heisenberg's results intelligible only on the grounds of Schrödinger's (1927a) interpretation:[23]

> The Heisenberg matrix elements q_ℓ^{km} come into the coefficients in such a manner that their co-operating influence on the intensity and polarisation of the part of the radiation concerned is completely intelligible on the grounds of classical electrodynamics. (p. 59)

This explanation of charge distribution and the intensity and polarization of spectral lines in terms of the spatial distribution of the wave function might have supplied the physical interpretation needed to make wave mechanics more basic than matrix mechanics. Unfortunately, as Schrödinger realized, the interpretation still labored under three difficulties. First, only the real part of $\psi\partial\bar{\psi}/\partial t$ was used, while the imaginary part was arbitrarily discarded. Either both parts should be used or some justification should be given for discarding the imaginary part. Second, if the physical interpretation really fits the mathematical formalism, then the integration of (11) over all space should give a finite value independent of the time. The reason for this is that the integration gives, apart from normalizing constants, the total charge, a conserved quantity. By virtue of (8), however, the integration over all space gives zero rather than a finite value. Third, the Schrödinger equation uses configuration space rather than real space. This basic difficulty with the Schrödinger interpretation requires a more detailed account.

The Schrödinger equation for a single particle in a central field of potential $V_c(x,y,z)$ is

$$(13) \qquad [-\hbar^2/2m\nabla^2 + V_c(x,y,z) - E_1]\psi(x,y,z) = 0 ,$$

where ∇^2 is $\partial^2/\partial x^2 + \partial^2/\partial y^2 + \partial^2/\partial z^2$. How should (13) be expanded to cover two particles in a central field? If $\psi(x,y,z)$ represents something, such as matter waves or charge density, distributed over real space, then the equation for two particles must be a function of the real spatial coordinates, x, y, z. To get such an equation in its simplest form, first consider the second particle separately. It too would be described by (13), but with E_2 for E_1. If, as Schrödinger suggested, this wave function is to be interpreted as a charge density $\rho_2(x,y,z) = e_2\int\bar{\psi}\psi\, dV$, then this distributed charge would interact with the distributed charge that is 'particle 1' by the quantum equivalent of the Coulomb interaction. To a first approximation this would be $V_{int} = e_1^2\int\rho_2(x',y',z')/(\underline{r}_1 - \underline{r}_2)\, dV'$. Therefore, if the proposed interpretation of the ψ-function is correct, then to a first approximation the Schrödinger equation for two particles should be[24]

$$[(-\hbar^2/2m)\nabla^2 + V_c + e^2 \cdot \int \frac{\bar{\psi}(x',y',z')\psi(x',y',z')}{|\underline{r} - \underline{r}'|} dv' - E]\psi(x,y,z) = 0 , \tag{14}$$

where

$$|\underline{r} - \underline{r}'| = \sqrt{(x - x')^2 + (y - y')^2 + (z - z')^2} .$$

The primes in (14) do not indicate different spatial coordinates but simply serve to keep the integration in order.

Equation (14) is a difficult integro-differential equation that can, in general, be solved only by approximation methods. However, there is an even more fundamental problem than calculational difficulties; the equation is nonlinear. For a nonlinear field, the basic principle of superposition does not hold. The field due to two particles is not simply the sum of the fields due to each particle separately.

In wave mechanics as Schrödinger developed it, one does not treat the interaction of two particles in a central field by using (14). Rather, one uses x_1, y_1, z_1 as the spatial coordinates of the first particle and x_2, y_2, z_2 as the spatial coordinates of the second particle and then combines two separate Schrödinger equations of type (13) plus an interaction potential to get

$$[-\hbar^2/2m(\nabla_1^2 + \nabla_2^2) + V_1(x_1,y_1,z_1) + V_2(x_2,y_2,z_2) + V'(x_1,y_1,z_1;x_2,y_2,z_2) - E]\psi = 0 . \tag{15}$$

Equation (15) is linear and can be solved by various approximation methods. The ψ-function that is the solution of this equation, however, can no longer be interpreted in any straightforward way as giving a distribution in real space. For any given particle i in an n-particle system, the triplet (x_i,y_i,z_i) functions as the independent variables on which the dependent variable ψ depends. Real and configuration spaces are easily confused in one particle system, since each uses the triplet (x,y,z). But for an n-particle system, configuration space is specified by the 3n coordinates $(x_1,y_1,z_1,x_2,y_2,z_2,\ldots,x_n,y_n,z_n)$, while real space is simply specified by the triplet (x,y,z) designating the location of spatial points relative to the chosen origin. The ψ-function is a function in configuration space, not real space. This presented not only a formidable difficulty for interpretation but also an abiding source of conceptual confusion.[25]

IV. PERTURBATION THEORY

The third communication in the quantization series (Schrödinger, 1926d) was received on May 10, 1926. In it Shrödinger treated time-independent perturbation theory. By adapting some methods Lord Rayleigh had developed in acoustics and utilizing the mathematical tools that had just become available through the monumental work of Courant and Hilbert, Methoden der mathematischen Physik, Schrödinger was able to give a development of time-independent perturbation theory that was both general and rigorous. To illustrate the scope of this method he gave a very detailed treatment of the Stark effect including line intensities, selection rules, and polarization rules, compared his results with the best data available, and even predicted the results of some special cases that had not yet been tested. Next to the derivation of the hydrogen spectrum this was the most resounding success of the new theory and conclusively established the efficacy of the general formalism that Schrödinger had developed.

The basic outline of this paper was probably planned before Schrödinger realized the need to work out a physical interpretation different from Heisenberg's. The perturbation theory he developed did not depend on differences between the two interpretations, but Schrödinger found one key point that did seem to differentiate the two interpretations. In matrix mechanics the generalized position coordinate is represented by a square matrix. Schrödinger could use his ψ-function, together with the density function, to determine an arbitrary component $q^{rr'}$ of this matrix (Schrödinger, 1927a, 82ff.).

$$(16)\quad q^{rr'} = \frac{\int q\rho(x)\psi_r(x)\psi_{r'}(x')\,dx}{\{\int\rho(x)[\psi_r(x)]^2dx\int\rho(x)[\psi_{r'}(x)]^2dx\}^{1/2}}$$

The interpretation of equation (16) supplied Schrödinger with the opportunity to distinguish between his and Heisenberg's interpretations of quantum theory. For Heisenberg, as Schrödinger interpreted his position, the square of (16) should be a measure of the probability of a transition from the rth to the r'^{th} electronic state. More precisely, it is a measure of the radiation resulting from this transition that is polarized in the q direction. Where Heisenberg's interpretation utilized the idea of quantum jumps and transition probabilities, Schrödinger was able to give this matrix element a simple physical interpretation that relied on continuity rather than quantum jumps. According to Schrödinger's interpretation, $q^{rr'}$ is a particular component of the amplitude of the periodically oscillating electrical moment of the atom. Consequently, the matrix can be interpreted as a kind of Fourier analysis, but one that uses the actual frequencies of emission rather than the unobserved harmonic overtones given by a straightforward Fourier analysis. Schrödinger (1926d) brought out the physical significance of this:[26]

> However, the idea of wave mechanics is not that of a sudden transition from one state of vibration to another, but according to it, the partial moment concerned--as I will briefly name it--arises from the <u>simultaneous existence</u> of the two proper vibrations, and lasts just as long as both are excited together. (p. 83)

The idea that characteristic lines in spectral radiation are due to the simultaneous existence of vibrational states certainly distinguished Schrödinger's position from the older interpretation, stemming from Bohr and taken over by Heisenberg, that radiation is due to a quantum transition from one state to another. The fate of this novel idea will be considered when we discuss Q4. The further development of Schrödinger's interpretation seems to have been stimulated, at least in part, by a letter H. A. Lorentz sent him on May 27, 1926.

Since the first Solvay conference in 1911, chaired by Lorentz, and his retirement from routine teaching in 1921, Lorentz had unofficially but effectively assumed a position as a sort of coordinator of the European physics community. When Schrödinger requested an evaluation of his new wave mechanics, Lorentz's generally commendatory reply carefully pointed out the outstanding difficulties in the physical interpretation Schrödinger was suggesting.[27] The first three objections Lorentz brought up concerned difficulties Schrödinger had already considered. The wave interpretation works for three-dimensional space but cannot be given a straightforward physical interpretation for multidimensional configuration space. Second, it is difficult to interpret particles as wave packets because wave packets spread. Third, there are some phenomena, such as the photoelectric effect, where electrons seem to act as clean particles without manifesting wave properties. Lorentz's final objection related more directly to the new interpretation Schrödinger was attempting to work out. Lorentz argued that the interpretation of radiation as beats, or combination tones of energy levels that are simultaneously excited, did not seem plausible. In the Bohr theory one could produce a radiation $\nu_3 - \nu_2 = (1/h)(E_3 - E_2)$ by supplying an amount of energy $E_3 - E_1$ that raised the electron from the ground state E_1 to the excited state E_3, from which it could spontaneously decay to the state E_2. For this case, Schrödinger's new interpretation would require the simultaneous existence of the energy levels E_0 (some constant value) $+ E_2 + E_3 - E_1$. This requirement did not seem to be consonant with the observed energy transitions.

Schrödinger's answer, sent June 6, 1926, revealed the tentative nature of the interpretation he was developing. He thought that he could solve the problem of configuration by attaching physical significance to $\bar{\psi}\psi$ rather than to ψ, and by relating this product to the charge density. He still could not explain the fact that ψ was complex rather than real. Also, as he admitted, the spreading of the wave packet remained a problem. To handle this, Schrödinger was toying with the idea that the spreading might be admitted and could be acceptable provided one is willing to sacrifice the doctrine that electrons can exist and preserve their identity in free space.

While these suggestions were somewhat tentative, Schrödinger was much stronger in his interpretation of the relation between simultaneous radiations within the atom and the energy radiated. He answered Lorentz's final objection by denying the identification of proper vibrations within the atom with Bohr energy levels. This, Schrödinger argued, would vitiate the correspondence principle transition from quantum to classical physics. Rather than consider the Bohr orbits as something basic, one should consider an individual orbit, at least for large quantum numbers, as built up from a superposition of very many proper oscillations.

This is a rather strange argument. The first real justification for the concept of matter waves was de Broglie's argument that Bohr's stationary orbits should be explained as the result of an integral number of wave nodes. Though, as noted earlier, this argument had some fuzzy aspects, it required one wave rather than a superposition of many waves for each proper orbit. Schrödinger, however, was now willing to drop de Broglie's duality almost completely and think of both electrons and radiation as nothing but complicated wave phenomena. Since he liked the idea that a free electron was a wave packet, it probably seemed appropriate to think of bound electrons as superpositions of waves, rather than individual waves. In this case, as Schrödinger noted, there must be a harmony between the vibration frequencies that effectively constitute the presence of electron orbits and the radiation emitted or absorbed. Even though this new view was not yet worked out in sufficient detail, Schrödinger believed it to be quite superior to the established Bohr view:[28]

> I was so extremely happy, first of all, to have arrived at a picture in which at least something or other really takes place with that frequency which we observe in the emitted light that, with the rushing breath of a hunted fugitive, I fell upon this something in the form in which it immediately offered itself, namely as the amplitudes periodically rising and falling with the beat frequency. By this I only meant: there is a *conceivable* mechanism by means of which these rising and falling amplitudes can excite light of *equal* frequency. The frequency discrepancy in the Bohr model, on the other hand, seems to me (and has indeed seemed to me since 1914) to be something so *monstrous*, that I should like to characterize the excitation of light in this way as really almost *inconceivable*. (Przibram, 1967, p. 61; italics in original)

The final paper in Schrödinger's (1926e) quantization series was received on June 21, 1926, some two weeks after the letter to Lorentz we have been considering. In it Schrödinger presented the time-dependent Schrödinger equation (though not in the full generality Dirac later gave it), time-dependent perturbation theory, and a relativistic wave equation equivalent to the Klein-Gordon equation. These achievements, significant as they are, are subordinated to the special emphasis characterizing this paper, the attempt to work out the promised new interpretation of wave mechanics in a way that comes to grips with the difficulties we have been considering. In developing this interpretation, Schrödinger was particularly concerned with showing how his theory yields explanations that are different from and superior to those given by matrix mechanics. These are the aspects of Q4 that we will consider here.

The basic interpretative problem confronting Schrödinger was the one present from the beginning--to explain the physical significance of the ψ-function. It is convenient to consider this problem in two stages: the ψ-function for a free electron and the ψ-function for a bound electron. Though Schrödinger was favorably disposed to the idea that free electrons may be considered as wave packets, he still would not make this a basis for the quantization series but relegated it to his later paper on the Compton effect. In Q4 he focused on the physical significance to be accorded the ψ-function <u>within</u> the atom, especially the multi-electron atom.

The basis of Schrödinger's interpretation is the idea, previously developed for the one-electron case, that $e\psi\bar{\psi}$ represents the charge density. This is extended to the multi-electron case by a five-step procedure of

(a) selecting some particular particle,
(b) keeping the triplet of coordinates that describe its position in ordinary mechanics fixed (this triplet [x,y,z] assigns a location in real space),
(c) integrating $e\psi\bar{\psi}$ over all the rest of the coordinates of the system (this is an integration in a configuration space of 3n - 3 dimensions for an n-particle system),
(d) multiplying by e,
(e) and then doing the same for every other particle.

After giving these rules, Schrödinger (1927a) summarized the interpretation they supported.

> This rule is equivalent to the following conception, which allows the true meaning of ψ to stand out more clearly. $\psi\bar{\psi}$ is the kind of _weight-function_ in the system's configuration space. The wave-mechanical configuration of the system is a _superposition_ of many, strictly speaking of _all_ point mechanical configurations kinematically possible. (p. 120; italics in original)

This interpretation appeared to solve the problems presented by configuration space. The product $e\psi\bar{\psi}$, when summed for the atom as a whole, is physically real and corresponds to the spatial distribution of electrical density. However, one can only get at this total result by a series of partial calculations. These intermediate steps require a detour through a physically unreal configuration space. Unfortunately, this solution though suggestive could not be considered final, for the Schrödinger equation still led to a complex, rather than a real, ψ-function. The proposed interpretation required a real, rather than a complex or imaginary, wave function.

The best that Schrödinger could do with this difficulty was to suggest a plausible hypothesis. A complex wave-function can always be written in terms of sines and cosines by the formula $e^{i\varphi} = \cos\varphi + i\sin\varphi$. It may be that one is really dealing with a real function, cos φ, and its time derivative, −(sin φ). If the total wave function of the atomic system can only be approximated through the five steps summarized earlier, then it seems reasonable to suppose that there is a real wave equation, probably of fourth order, to which Schrödinger's present equations are just first approximations. This true wave equation should have functions that are real rather than complex.

This interpretation remained tentative and incomplete, yet it supplied the only basis Schrödinger could develop for showing the superiority of his system to Heisenberg's. To bring this out, Schrödinger tried, in effect, to beat Heisenberg at his own game. The justification of the Ladenburg-Kramers dispersion formula had played a crucial role in the development that led Heisenberg to quantum mechanics.[29] Schrödinger now presented an interpretation which, he believed, could make the formal results obtained by Kramers and Heisenberg physically intelligible. Earlier we considered the way Schrödinger used the ψ-function, and the electromagnetic interpretation he accorded it, to explain the dipole moment of an atom subject to electromagnetic radiation. Now these considerations were extended to the case where two free vibrations are excited, u_k and u_ℓ. In addition to the sum of these two terms, Schrödinger expected interaction terms between the forced vibrations of u_k and the free vibrations of u_ℓ and vice versa. The frequency of these interaction terms is not ν, the frequency of the incident radiation, but

$$|\nu \pm (E_k - E_\ell)/h| \tag{17}$$

The Kramers-Heisenberg formula, developed prior to quantum mechanics, had this term, as did the Born-Heisenberg-Jordan paper on quantum mechanics. But the Heisenberg approach, Schrödinger insisted, could explain neither the physical significance of this term nor the conditions for its experimental verification.

The conditions for experimental verification are, he argued, uniquely given by Schrödinger's electromagnetic interpretation of the ψ-function. In order for this postulated, as yet unobserved, radiation to occur, both states u_k and u_ℓ must be strongly excited in the same individual atom. For such radiation to be detectable, such simultaneous excitations must occur in many atoms. The interpretation of this difference term in the frequency as due to the simultaneous excitation of different states, or vibrational levels, is peculiar to Schrödinger's interpretation of wave mechanics. Matrix mechanics, which abandons space-time descriptions, has no way of even expressing this. Schrödinger (1927a) made this difference quite explicit:

> As far as I can see, the above-mentioned dispersion theory of Heisenberg, Born and Jordan does *not* allow of such reflections as we have just made, in spite of the great formal similarity to the present one. For it only considers *one* way in which the atom reacts to incident radiation. It conceives the atom as a timeless entity, and up till now is not able to express in its language the undoubted fact that the atom can be in *different* states at different times, and thus, as has been proved, reacts in different ways to incident radiation. (pp. 111-112; italics in original)

V. THE DEMISE OF THE WAVE INTERPRETATION

With the completion of Q4, the most positive and productive period in Schrödinger's career came to a close. The next few papers he wrote were essentially attempts to revise, repair, and extend his new interpretation and to answer the objections that had been brought against it. For this reason these papers will be treated more briefly than their predecessors. Schrödinger's final attempt to develop an adequate physical interpretation of the ψ-function had three distinctive aspects: a detailed development of the hypothesis that $e\psi\bar{\psi}$ represents the spatial distribution of charge density, the suggestion that free electrons might really be wave packets, and the account of the predicted but not yet observed extra frequency in (17) as the result of the simultaneous excitation of two or more vibrational states within the atom.[30] We will consider the future development of these points in more or less chronological order.

The strongest support for the electromagnetic interpretation of $e\psi\bar{\psi}$ came from the proof that it fulfilled a continuity equation and therefore seemed to account for charge conservation. However, even in Schrödinger's original proof,[31] the inclusion of the charge e played no real role. This may be most easily seen by a simple argument familiar to physicists. Consider the wave equations for a ψ-function and for its complex conjugate $\bar{\psi}$

(18) $$(-\hbar^2/2m\nabla^2 + V)\psi = i\hbar\partial\psi/\partial t \ ,$$

(19) $$(-\hbar^2/2m\nabla^2 + V)\bar{\psi} = -i\hbar\partial\psi/\partial t \ .$$

By multiplying (18) on the left by $\bar{\psi}$, multiplying (19) on the right by ψ, subtracting (19) from (18), and rearranging the terms, one obtains

(20) $$-\hbar/2im \ \underline{\nabla}(\bar{\psi}\underline{\nabla}\psi - \nabla\bar{\psi}\psi) = \partial/\partial t(\bar{\psi}\psi) \ .$$

Equation (20) has the form of a continuity equation,

(21) $$\partial\rho/\partial t + \underline{\nabla}\cdot\underline{S} = 0 \ ,$$

in which

(22) $$\rho = \bar{\psi}\psi$$

and

$$(23)\qquad \underline{S} = \hbar/2im(\bar{\psi}\underline{\nabla}\psi - \underline{\nabla}\bar{\psi}\psi) = \hbar/mI_m(\bar{\psi}\underline{\nabla}\psi) .$$

If equations (22) and (23) are multiplied by e, then ρ may be interpreted as a charge distribution and $\underline{S}$ as a current. Schrödinger had essentially the same equations in his expression $d/dt \int \psi\bar{\psi}\rho \, dx = 0$, where ρ is a density factor. The electromagnetic interpretation of these equations seemed uniquely probable as long as there were no competing interpretations.

As is well known, Max Born soon produced the probability interpretation of the quadratic factor $\psi\bar{\psi}$ in his theory of collisions, and this interpretation, which Pauli generalized, won immediate and lasting acceptance.[32] This interpretation preserved what was valid in (21), the conservation of probability, or in a statistical interpretation, the conservation of the number of particles. In doing this it also eliminated the most untenable feature of Schrödinger's electromagnetic interpretation, the inconsistency between the indefinite spreading of the wave function and the experimental fact that the electron retains its integral value and relatively sharp localization.

In late summer of 1926, Niels Bohr invited Schrödinger to Copenhagen to discuss the interpretation of quantum theory. The anecdotal features of their clash have been repeated and elaborated until they have become part of the established lore of modern physics.[33] For our purposes, however, the content of their disagreement is of more concern than its emotional aspects. Bohr accepted the mathematical formalism that Schrödinger had developed and agreed with Schrödinger that the wave properties of matter had to be included in any acceptable interpretation, but he rejected the physical interpretation Schrödinger had developed on the grounds that it did not account for the corpuscular aspects of matter and could not even explain Planck's treatment of black-body radiation. This, like the old quantum theory that sprang from it, required discrete energy states and quantum jumps between states. Ultimately the only reply the beleaguered Schrödinger could give was: "If all this damned quantum jumping were really here to stay, I should be sorry I ever got involved with quantum theory" (Heisenberg, 1971, p. 75).

Heisenberg's account reflects Bohr's side of the argument. Schrödinger's own reaction is best seen from a letter he wrote to Bohr on October 23, 1926,[34] a few weeks after leaving Copenhagen, in which he explained why he could not accept the doctrine of complementarity which Bohr was then developing and which we will treat in the third article in this series.

> Here there apparently exists a contradiction and you say: our earlier words and concepts do not extend this far. I can not be satisfied with this position and I cannot, accordingly, deduce for myself the right to operate any further with contradictory expressions.

Then Schrödinger contrasted this summary of Bohr's position with his own ideas on the goal of scientific explanation.

> What hovers before my eyes is only a single thesis: even if a hundred attempts miscarry, one would not give up the hope of reaching the goal of a representation of the true properties of space-time events through--I do not say classical pictures--but through representations that are free of logical contradictions. It is extremely probable that this is possible.

Away from the competitive strife of Copenhagen, Schrödinger tried to work out this program in the privacy he found necessary for sustained thought. As mentioned in the letter to Bohr, he focused on the space-time representation of events rather than on the configuration space formulation and the physical interpretation that he accorded it and that Bohr had so severely criticized. The first two articles (1927c, 1927d) he wrote on his return from Copenhagen reflect an effort to revitalize de Broglie's idea of material waves in real space, rather than Schrödinger's own configuration space waves. Many physicists independently, and approximately simultaneously, had developed the relativistic wave equation now known as the Klein-Gordon equation.[35] Though Schrödinger was one of the codiscoverers, he used Gordon's development of relativistic quantum mechanics as the basis for his new interpretation. Gordon (1926) gave a relativistically invariant four-dimensional expression of the continuity equation that supplied the justification for Schrödinger's electromagnetic interpretation of the ψ-function.[36] For a short time, Schrödinger (1927a) seems to have banked rather heavily on the acceptability of this new basis:

> The Hamiltonian principle from which the exact relativistic differential equation for de Broglie waves can be obtained, appears to justify completely the hopes which I had set upon an intimate blending together of wave mechanics and classical electrodynamics. (p. 129)

This new interpretative basis, unfortunately, had even more flaws than the older one based on nonrelativistic configuration-space wave functions. It did not yield the established relativistic corrections for hydrogen. Any attempt to extend the equation from one to many particles precipitated the problems considered earlier of a choice between de Broglie wave functions in real space and Schrödinger wave functions in configuration space. In his treatment of the Compton effect, Schrödinger blurred this distinction by focusing on a steady-state case where only the density distribution need be considered. A few months later, Dirac (1926) gave the same problem a more rigorous quantum mechanical treatment that did not involve such simplified assumptions of a steady state.[37] Schrödinger adapted his approach to classical electromagnetic theory, coupling ψ-waves to Maxwell fields. Here again, Dirac (1927b) provided the necessary corrective, one that Schrödinger soon accepted, by substituting a quantized electromagnetic field for the classical Maxwell field. Finally, the new interpretation was no more able to solve the problem of dispersion of wave packets than was the older one.

In the background behind these attempts at interpretation and reinterpretation there was still the competition between Heisenberg and Schrödinger. As indicated earlier, Heisenberg initially reacted strongly against wave mechanics considered as a competing system. However, he quickly accepted the proofs developed by Schrödinger and Pauli that the two systems were mathematically equivalent. From this time on, Heisenberg (1926a) carefully distinguished between Schrödinger's mathematical method, which Heisenberg used to supplement matrix methods, and the physical interpretation Schrödinger attached to this method.[38] In the first paper he wrote utilizing both methods, Heisenberg argued that his own interpretation was preferable to Schrödinger's because it treated the many-body problem as a collection of particles in real space rather than as waves in configuration space. However, the intuitive content of his own theory, Heisenberg had to admit, was not yet clear. The inclusion of Bose-Einstein statistics led to a special solution of the many-body problem that could be interpreted as phase relations between the partial systems or particles. This seemed to accord better with a wave than with a particle interpretation.

From this time on, Heisenberg accepted Schrödinger's method and used it to supplement quantum mechanics or even to replace it in solving problems that were forbiddingly difficult by the methods of matrix mechanics, such as the spectra of atomic systems with two electrons.[39] Yet Heisenberg remained unwilling to accept Schrödinger's interpretation of this formalism. Through the fall and winter of 1926 this unwillingness became the focus of a growing rift between Bohr and Heisenberg. During this period Heisenberg wrote two papers that clearly manifest an attempt to show quantum mechanics superior to wave mechanics as a physical interpretation and to counter Schrödinger's criticism that quantum mechanics remained too unintuitive to be acceptable. The first paper (Heisenberg, 1926b)--a qualitative, nontechnical survey of the new developments in quantum theory--carried the now familiar distinction between Schrödinger's mathematical formalism, which Heisenberg found acceptable, and the physical interpretation, which presented more formidable problems. Here Heisenberg clearly distinguished between de Broglie waves in real space and Schrödinger waves in configuration space. He found Schrödinger's interpretation of the latter unacceptable on the grounds that the assumption of continuity did not seem in accord with such discontinuous phenomena as black-body radiation and dispersion. This resolution of competing interpretations, however, was still far from conclusive. Heisenberg was forced to admit that the particle interpretation was ambiguous in that the theory did not allow the specification of a particle's position as a function of time nor did it easily lend itself to an explanation of the interference effects that seemed analogous to interference effects in light.

Fluctuation phenomena, treated in the second paper (Heisenberg, 1927a), were seen as having a special significance for bringing out the intuitive consequences of discontinuity in processes occurring in very limited space-time regions. He considered two similar atoms, a and b, having energy states, E_n and E_m. When these are coupled together, the interaction between them can be viewed in two ways--as regularly occurring energy jumps in each atom due to the exchange of photons between them or as something analogous to the classical interaction between two coupled oscillators. In the latter interpretation the energy exchange is continuous, while in the former it is discontinuous. The two interpretations lead to the same conclusions for all observable effects, chiefly because of the impossibility of measuring individual interactions in such a way as to settle the continuity-discontinuity issue. However, the continuity interpretation cannot be accorded any physical significance except in cases where the de Broglie wave length of the particles is small compared to the size of the interaction volume. Heisenberg interpreted this as favoring discontinuous energy jumps. But here again, as he clearly realized, his argument was far from conclusive. Some three weeks after Heisenberg finished his paper, Jordan (1927) submitted a paper on essentially the same problem. In it he proved that both quantum mechanics, with the assumption of energy discontinuities, and wave mechanics, which assumes a continuous energy exchange between the two coupled atoms, agree in all empirically testable consequences.

To render the dialectics of this development intelligible, it is necessary to disrupt the chronological order of the published papers. Heisenberg's indeterminacy paper was written prior to the Schrödinger paper we are about to consider. Schrödinger's paper, however, answered the Heisenberg papers we have just summarized, as well as the objections Bohr had raised, but made no reference to the indeterminacy paper, which was not yet available at the time Schrödinger completed his article.[40]

In his final sustained attempt to salvage the wave interpretation, Schrödinger (1927b) abandoned his efforts to revise de Broglie real-space waves and returned to his own configuration space waves. Following Heisenberg and

Jordan, he considered two interacting systems: I, with energy levels E_k and $E_{k'}$, and II, with energy levels E_ℓ and $E_{\ell'}$. A necessary condition for a quantum interaction is that $E_k - E_{k'} = E_\ell - E_{\ell'}$. Heisenberg had interpreted this in terms of energy jumps. Schrödinger (1927a) reworked the problem and obtained equivalent results without introducing any assumptions of discrete energy levels and quantum exchanges of energy. If, he argued, frequency is considered basic rather than energy, then the exchange interaction could be interpreted as a resonance phenomenon like the sympathetic oscillations of coupled pendulums. If the basic postulate of energy jumps plays no role, then, in accord with Occam's razor, it should be discarded.

> According to the fundamental principles of research which are commonly regarded as correct, does not the foregoing compel us to exercise the utmost caution with respect to the quantum postulates even (I almost feel inclined to say) to distrust them--quite apart from their axiomatic unintelligibility. (p. 142)

By extending these considerations to the interaction of a system, which could be called a thermometer, with an extremely large system having a very dense eigenvalue spectrum, or a heat bath, Schrödinger was able to deduce Planck's formula from his wave mechanics, thus answering Bohr's challenge. In passing, he also showed that this interpretation yielded Bohr's probability account, but he accorded it a derivative rather than a primary status. Accordingly, the essential aspects of Schrödinger's interpretation could be salvaged if frequency was accepted as a fundamental physical concept and energy accorded the derivative status yielded by the formula $\nu_{nm} = (E_n - E_m)/h$--a heroic attempt, but one that soon fell victim to Heisenberg's indeterminacy principle.

Heisenberg's indeterminacy principle has emerged from the flux of science in progress to become one of the most philosophically pregnant limiting principles established by modern science. Yet it is only by resituating this principle in the flux from which it emerged that we can see the decisive role Heisenberg's paper played in settling the struggle of competing interpretations. Fortunately, Heisenberg has supplied the background material necessary to understand the genesis of this paper.[41]

In the fall of 1926 and on through the winter, Bohr and Heisenberg kept discussing the problem of interpreting quantum theory. Though they worked well together on other issues, they were unable to reach any accord on Bohr's contention that the wave interpretation, in spite of its shortcomings, must have some validity. As Heisenberg later explained it (Kuhn et al., 1967, interview 8):

> The difficulties in the discussion between Bohr and myself was that I wanted to start entirely from the mathematical scheme of quantum mechanics and use Schrödinger theory perhaps as a mathematical tool sometimes, but never enter into Schrödinger's interpretation which I couldn't believe. Bohr, however, wanted to take the interpretation in some way very seriously. (pp. 14-15)

The disagreement over the acceptability, or salvageability, of the Schrödinger interpretation became so heated that Bohr forced Heisenberg to break down in tears. Eventually, the two men were no longer able to work together. Bohr left to go skiing in Norway and to think through his own position, while Heisenberg remained in Copenhagen writing up his views in a long letter to Pauli, hoping that Bohr would take Heisenberg's position seriously if Pauli found it acceptable. This letter to Pauli was effectively the first draft of the indeterminacy paper.

The strongest criticism that Schrödinger had leveled against matrix mechanics was that the theory was so unintuitive that it was hardly acceptable as a physical theory. Heisenberg's paper (1927b) "On the Intuitive Content of Quantum-theoretical Kinematics and Mechanics"[42] was intended, inter alia, to answer this criticism. The paper begins with a brief analysis of the notion of intuitive understanding. We believe that we have an intuitive understanding of a theory when we know in a qualitative way the experimental consequences of this theory for all simple cases and when we are assured that it is free from contradictions.[43] Using this rather operationalist account of intuition, Heisenberg went on to explain the significance quantum theory could accord to such concepts as 'location of an electron', 'path of an electron', 'velocity', and 'energy'. By applying these considerations to such ideal experiments as the gamma-ray microscope, Heisenberg derived the indeterminacy principle in the form $p_1 q_1 \sim h$, where p_1 is the degree of exactness in momentum and q_1 that for position. A more formal approach, based on the Dirac-Jordan transformation theory, showed that the minimal uncertainty for the simple case of a Gaussian wave packet is $p_1 q_1 \sim h/2\pi$.

The Schrödinger wave function, Heisenberg argued, admits of the same interpretation. Such an interpretation, in fact, brings out the deep significance of the linearity of the solution. A superposition of solutions is also a solution. There is no prospect of getting around this difficulty by reliance on nonlinear relativistic equations (Heisenberg, 1927b).[44] Though many of his readers might have missed the significance of this reference to nonlinear relativistic wave equations, Schrödinger could not. The interpretation of ψ-waves as relativistic de Broglie waves in real space, an interpretation that Schrödinger had defended in two papers previously considered, required nonlinear equations in applications to two-electron atoms.

Against the backdrop of these theoretical considerations, Heisenberg examined Schrödinger's contention that it is possible to consider particles as wave packets. First he considered free particles and argued that the wave packet that Schrödinger had set up for the one-dimensional simple harmonic oscillator does not spread in the course of time only because the Fourier overtones are multiples of the fundamental frequency, something not true in other cases. A more precise treatment of the wave-packet representation of a particle can be had by applying transformation theory to the linear motion of a mass-point. Suppose that, at the time, the position is measured with a degree of exactness q_1 and the momentum with a degree of exactness p_1 (where $p_1 q_1 = h/2\pi$). The form of the transformation function from a representation in which q_1 is arbitrarily small to a representation in which the position matrix is diagonal shows that future positions are not determined by the theory with any precision at all. More concretely, one can fix the position of an electron at a given time by hitting it with a gamma ray of very short wavelength. But then the high Compton recoil makes future positions quite unspecifiable. The wave packets that relate to measurable information have the same limitations. Accordingly, they cannot be interpreted as waves in real space. Rather, they should be interpreted as probability packets specifying relations between past measurements and possible future measurements.

Heisenberg examined the further idea that wave packets can represent bound electrons by using the same double process of arguing both from theoretical considerations and from an examination of idealized measurements. Suppose an electron is in a very excited state, for example, n = 1000. Then the path could, in principle, be observed with fairly long wavelengths of light. While this light would not have the energy to knock the electron out of the atom, it would impart sufficient recoil to displace the electron to some other orbit, for example, between n = 950 and n = 1050. Since the precise orbit proper to an excited state cannot be known with greater accuracy than this, one would have to represent the electron's orbit by a superposition of wave packets proper to these hundred orbits. Again, this is really a probability packet and one that disperses in time, though it remains within the atom.

Schrödinger had interpreted the radiation emitted or absorbed by atoms in terms of the simultaneous excitation of excited states. By a detailed examination of different ideal experiments, Heisenberg showed that this notion has no operational significance. In resonance radiation, for example, atoms bombarded with a frequency exactly corresponding to a difference in energy levels $\nu_{12} = (E_2 - E_1)/h$ will reradiate coherent radiation. Because of this coherence, the phase can be determined precisely. However, because this is a cooperative phenomenon, there is no way of knowing which atoms are in state 1 and which in state 2. There are ways in which atoms in different energy states can be discriminated, such as passing the atoms through an inhomogeneous magnetic field. But any field that could separate atoms according to their energy states would also destroy the coherence of the radiation and with it any knowledge of phases. It is, accordingly, physically meaningless to give any precise description of the behavior of electrons in atoms, regardless of whether the description is couched in wave or particle language.

After showing in detail that Schrödinger's interpretation could not be given any operational significance, Heisenberg turned the charge of being unintuitive back against Schrödinger. Quantum mechanics, as interpreted by Heisenberg, is quite intuitive in the sense that the indeterminacy principle allows one to think through the experimental consequences of quantum theory in all simple cases. Schrödinger's wave mechanics is certainly more intuitive in the sense of having more manageable mathematical forms. In the question of physical principles, however, the popular intuitiveness of wave mechanics must be considered a deviation from the straight path (Heisenberg, 1927b, p. 196).

After this rather devastating criticism, Heisenberg concluded with a rather surprising note, added after the text was submitted for publication, indicating that the wave and particles pictures could be considered complementary. This note, however, was really more a peace offering to Bohr than a concession to Schrödinger. After Heisenberg finished his text, Pauli came to Copenhagen and mediated the disagreements between Bohr and Heisenberg. Through this note, Heisenberg acknowledged the compatibility of his views with Bohr's, a position he soon came to affirm much more emphatically.

Shortly after Heisenberg showed the particles could not be interpreted as nothing but wave packets, Paul Ehrenfest (1927) showed how a concept of a wave packet could serve as a bridge between classical and quantum mechanics. He proved that, regardless of spreading, the center of mass of a wave packet obeys the quantum equivalent of Newton's second law of motion: The expectation value of the time derivative of the momentum is equal to the negative gradient of the potential energy function, which may be called a force.

I have been unable to find any source material documenting Schrödinger's immediate reaction to Heisenberg's paper, yet this very lacuna may be revealing. After the paper that we have summarized as a last attempt to salvage the wave interpretation, Schrödinger wrote no technical papers for a year and a half.[45] This dearth, following upon his remarkable productivity, would seem to indicate a period of deep discouragement. At the Solvay conference, held in Brussels October 24-29, 1927, Schrödinger simply summarized the essentials of wave mechanics and presented his electromagnetic interpretation with no new refinements. In answer to a question de Broglie raised, Schrödinger (Solvay Conference, 1928) discussed the introduction of the wave-packet idea and concluded:[46]

> It has since been found that the naive identification of an electron, travelling over a macroscopic orbit, with a wave-packet encounters difficulties and can not be accepted as literally true. The principal difficulty is this, that a wave packet is scattered in all directions when it encounters an obstacle, an atom for example. (p. 288)

He also indicated a willingness to go along with a practical, though theoretically unsatisfactory, resolution of the wave-particle duality: "The compromise proposed on different sides consists in admitting an association of waves and point-electrons; I accept it only as a way of resolving the difficulty."

It was not, as far as I know, until December 1928 that he first publicly discussed the indeterminacy principle. Then he indicated that he accepted the indeterminacy principle as valid but felt that it still allowed a choice of strategies (Schrödinger, 1957).

> One may believe either (1) that matter has really a wave structure. Then the uncertainty principle is an immediate consequence. Or (2) one may think that the uncertainty principle is the more fundamental. The wave theory then is simply an auxiliary construction for the convenience of grasping and representing the principle. (p. 161)

It seems that Schrödinger initially followed the first option. In March 1928, he delivered a series of lectures in London on wave mechanics, which were subsequently published in book form (Schrödinger, 1928). Though these lectures contain little physics beyond the Q1-Q4 series, which they summarize, they present a slightly more cautious interpretation of the wave model: "The statement that what really happens is correctly described by describing a wave-motion does not necessarily mean exactly the same thing as: what really exists is the wave motion" (p. 6).

In this book Schrödinger defends his interpretation of $e\psi\bar{\psi}$ as the spatial distribution of charge. The defense rests on two arguments that are already familiar. This interpretation allows an intelligible explanation of the emission and absorption of radiation. Here, as earlier, Schrödinger emphasized the predicted, but not yet observed, extra dispersion frequency $|(E_k - E_\ell)/h|$. This, Schrödinger argued, should be interpreted as the interaction of two frequencies that are simultaneously excited in the same atom. Second, he felt that an interpretation in which frequency, rather than energy, is basic removes the unintelligible discontinuities involved in the acceptance of quantum jumps. Thus his book concludes with the rhetorical question: "Is it quite certain that the conception of energy, indispensable as it is in macroscopic phenomena, has <u>any</u> other meaning in micro-mechanical phenomena than the number of vibrations in h seconds" (p. 53). For Schrödinger, as for Heisenberg, the attempt to defend an interpretation drove him into a form of conceptual analysis.

Within the next few months, advances in physics undercut both of the arguments Schrödinger relied on. The predicted extra dispersion frequency was detected experimentally by C. V. Raman (Raman & Krishnan, 1928) in India and by Landsberg and Mandelstamm (1928) in Russia. The nature of this radiation is worth considering. When a parallel beam of monochromatic light goes through a gas, a liquid, or a transparent solid, a small fraction of the light is scattered in all directions. If the spectrogram is prepared so that the scattered lines identical in frequency with the light source are strongly overexposed, some weak additional lines are found that do not appear in the spectrum of the light source. For a given scattering substance, the displacement of these lines from the line corresponding to the incident radiation is independent of the wave length of the radiation used. These lines have exactly the form predicted by equation (17).

The formula works perfectly; Schrödinger's interpretation of it does not. In a gas or a liquid, some of the extra lines can only be explained as induced rotations or vibrations of the molecules struck by the incident radiation.[47] The idea of a molecule having different rotational or vibrational states simultaneously has no plausibility at all. However, these results may easily be accommodated to the simple explanation given in Smekal's (1923) original prediction of this effect. If a molecule with velocity v and energy state E_k (whether electronic, vibrational, or rotational) collides with a photon of energy $h\nu$ and passes to a state of velocity v' and energy E_ℓ, then the conservation of energy yields

$$mv^2/2 + E_k + h\nu = mv'^2/2 + E_\ell + h\nu'$$

Since the velocity change is relatively negligible, $\nu' = \nu + (E_\ell - E_k)/h$. This argument not only yields the correct frequency; it also reflects the deeply entrenched custom of considering energy, rather than frequency, a basic concept and its conservation a basic principle.

In that same year, Dirac (1928a, 1928b) developed a relativistic wave equation that, unlike the Klein-Gordon equation, included spin and yielded the proper relativistic corrections for the hydrogen atom. This was an advance that Schrödinger accepted and eventually expanded upon. Though the negative energy states this relativistic wave equation yielded were not correctly interpreted until 1932, the equation itself requires quantum jumps between separated energy levels.

Years later, Schrödinger (1953) summarized the basic reasons that led him to abandon his original wave interpretation. Particles cannot be pictured as wave packets because wave packets disperse. The electronic interpretation of the ψ-function does not yield a self-consistent interpretation of the hydrogen atom. Planck's radiation law, and some nonequilibrium processes such as the photoelectric effect, requires discrete energy levels and quantum transitions between them. Though, as he admitted, the Copenhagen interpretation has an admirable self-consistency, he could never accept it as an adequate representation of atomic processes. In Schrödinger's view, quantum jumps could no more be a real part of nature than Ptolomaic epicycles. To the end, he remained a loner, outside the mainstream of quantum developments, severely critical of the theory he had done so much to shape.

Notes

I wish to thank Professor John Heilbron and Ms. Judy Fox for their cooperation in making available the material in the center for History of Science and Technology at the University of California, Berkeley. I also thank the directors of the Niels Bohr Institute in Copenhagen for allowing access to microfilm copies of Niels Bohr's correspondence. I am especially grateful to Professor Linda Wessels for making available prepublication copies of her own work on Schrödinger.

[1]This article is part of a series analyzing quantum mechanics as a conceptual revolution. The other articles

in the series are listed in the references. I hope to give a more adequate theoretical interpretation of this conceptual revolution in a later study.

[2]These articles are translated in Erwin Schrödinger, <u>Collected Papers on Wave Mechanics</u> (1927a). The German originals will be cited when we treat the individual articles.

[3]See, for example, the work of Jammer (1966), Hund (1974), or Gerber (1969).

[4]Linda Wessels (1976) gave a general interpretation of Schrödinger's development in her PhD dissertation.

[5]This is a point that I have developed in detail in MacKinnon (1977).

[6]The background to this may be found in Salpeter, Corson, and Bauer (1964).

[7]In an obituary notice on Schrödinger, Dirac claimed: "He immediately applied his method to the electron in the hydrogen atom, duly taking into account relativistic mechanics for the motion of the electron, as de Broglie had done. The results were not in agreement with observation. We know now that Schrödinger's method was quite correct, and the discrepancy was due solely to his not having taken the spin of the electron into account. However, electron spin was unknown at that time, and Schrödinger was very disappointed and concluded that his method was no good and abandoned it. Only after some months did he return to it, and then noticed that if he treated the electron non-relativistically his method gave results in agreement with observation in non-relativistic approximations." This notice is in Dirac (1961). This early relativistic work presents something of a puzzle. A former student of Schrödinger's, D. M. Denison, claimed that Schrödinger gave him a typed carbon copy of his unpublished early article on relativistic wave mechanics (in his third interview with Thomas Kuhn, on January 30, 1964; Note 1, folder 3, p. 7). Yet no copy of this paper remains nor is there any trace of this early work in the copious notebooks that Schrödinger left, at least in those available in microfilm copy in the Berkeley Center for the History of Science.

[8]The English translation (Schrödinger, 1927a) is misleading in citing the conditions ψ must fulfill. It reads "real over the whole of co-ordinate space" (p. 2)

where it should have "real over the whole of configuration space" as a translation of <u>Konfigurationenraum</u>. The significance of this distinction will be treated later.

[9]Because ψ tends to zero at infinity as $1/r$, $\partial\psi/\partial r$ tends to zero as $1/r^2$. Hence, the surface integral, involving a factor of r^2, is still of the same order at infinite as at finite distances. Allowing this would prohibit the continuous spectrum for positive total energy values. Accordingly, Schrödinger added the further postulate that the integral effectively vanishes over an infinite sphere.

[10]Pascual Jordan (1973, pp. 294-299) discusses David Hilbert's Göttingen seminar on the mathematical foundations of quantum theory in which almost all the contributors to quantum theory, except Schrödinger, participated. Hilbert was convinced on general grounds that there should be eigenvalue equations having both discrete and continuous solutions and was delighted when Schrödinger found a relatively simple example of such an equation.

[11]A description of the wave front as progressive but stationary may seem contradictory. It is really Schrödinger's abbreviated way of bringing out two distinctive features of the solution. His Hamilton-Jacobi equation yields a family of surfaces for the action function in configuration space. It is stationary in the sense that at any earlier or later instant than the comparison instant the same group of surfaces illustrates the same action-distribution. It is progressive in the sense that the action value for any particular point in configuration space is linearly dependent on time.

[12]Following Schrödinger's precedent (1927a, p. 45), we will use 'Heisenberg' in referring to the papers of Heisenberg, Born, and Jordan, as well as to work done by Heisenberg alone.

[13]The three-man paper (Heisenberg's term) is Born, Heisenberg, and Jordan (1926). Schrödinger (1927a, p. 47) used the relation between P_ℓ and $\partial/\partial q_\ell$ as his point of departure developing the relation between wave and matrix mechanics.

[14]This letter is given and its background explained in an article by van der Waerden entitled "From Matrix Mechanics and Wave Mechanics to Unified Quantum Mechanics" (1973, pp. 276-296). The letter is on pages 289-293 in

the German original and on pages 278-282 in the English translation.

[15]When he presented the relationship between wave and matrix mechanics (Wentzel, 1926, p. 522), he cited as sources Schrödinger's article and a private communication from Pauli.

[16]This interview took place on October 17, 1962. The citation is from Kuhn, Heilbron, Forman, and Allen (1967, folder 3, p. 22).

[17]A copy of this letter is available in the Berkeley Center for the History of Physics and had not yet been catalogued when I read it. I wish to thank Professor John Heilbron for bringing it to my attention.

[18]The background to this work is treated in Jammer (1966, pp. 275-276).

[19]Some of Hanson's statements should be revised in the light of later historical studies. A more precise indication of the scope and limitation of Schrödinger's proof may be found in van der Waerden (1973, pp. 226-227). These limitations are of a purely formal nature and were soon obviated by Dirac's unified theory of quantum mechanics. The quantum theorists of 1926 were justified in accepting the two theories as mathematically equivalent.

[20]Scott (1967, chap. 2) gives a good summary of Schrödinger's intellectual development prior to his development of quantum theory.

[21]This was translated into English in Schrödinger's _Science and the Human Temperament_. The unrevised reprint was retitled _Science, Theory and Man_ (1957). The complex scientific and political background to the then current controversy about causality may be found in Forman (1971).

[22]Also, in his first letter to Bohr sent on May 24, 1924 (Note 1, reel 16, sec. 2).

[23]I believe that the strongest case that could be made for an underlying continuity in Schrödinger's technical physics and his later opposition to the Copenhagen interpretation would come more from his use of the word _intelligible_ than from his philosophical reinterpretations of physics. The rationalist tradition, in its

variant forms, has always stressed the primary role of intelligible principles in any adequate explanation of reality and tended to reject discontinuities in nature on the grounds that they are unintelligible. The empirical tradition, on the other hand, usually tends to stress the primacy of observation and is willing to accept as basic whatever laws emerge from a process of scientific investigation, inductive generalization, and hypothetical-deductive reasoning, regardless of whether these laws have an independent intelligibility or must simply be accepted as given. This criterion would put Bohr, Heisenberg, Dirac, and Pauli in the empiricist camp, while Schrödinger and Einstein, at least in his later years, would be rationalists. Their somewhat nebulous pantheistic views would accord well with the theological orientation that has usually grounded the rationalist stress on intelligibility. Such arguments, however, tend to be post factum and rather unconvincing.

[24]This is discussed in some detail in Sin-Itiro Tomonaga (1966). He calls equation (14) a "de Broglie equation."

[25]Thus, Born and others have accused Schrödinger of confusing real and configuration space. I believe that these accusations have been adequately handled by Scott (1967, pp. 77-80) on the grounds that, though Schrödinger was well aware of the distinction, he avoided making it in popular lectures because of its technical nature.

[26]Zusammenbestehen (simultaneous existence) is italicized in the original text (Schrödinger, 1926d, p. 465).

[27]These letters may be found in K. Przibram (1967, pp. 43-54, Lorentz's first letter; pp. 55-61, Schrödinger's reply). The later letters contain more technical details but nothing new on the issues we are considering.

[28]This is a very interesting quote manifesting both Schrödinger's emotional attachment to his new interpretation and his unwillingness to present his new theory as dependent on this interpretation (only a conceivable mechanism). After completing the basic quantization series, he was willing to come out more clearly in favor of interpreting particles as wave packets. But he was careful to present this as a personal view rather than as the basis of his new theory. Thus, in an English summary

of his quantization series, he wrote: "The point of view taken here, which was first published in a series of German papers, is rather that material points consist of, or are nothing but, wave systems." He went on to say that though this point of view may be wrong, it seems preferable to the counterpart interpretation which stresses only particles. This is in his paper "An Undulatory Theory of the Mechanics of Atoms and Molecules" (1926, p. 1049).

[29]A detailed discussion of the role this assumption played in Heisenberg's development may be found in the source cited in MacKinnon (1977).

[30]A detailed account of Schrödinger's interpretation may be found in Scott (1967, chap. 4). Scott, however, does not try to give a chronological account of the changes in Schrödinger's interpretation or of the dialectic between Schrödinger's and opposed positions.

[31]This proof was given in Q4. See Schrödinger (1927a, pp. 120-123).

[32]Max Born first suggested this interpretation in a correction added in proof to his article "Zur Quantenmechanik der Stossvorgange" (1926; received June 25, 1926) and developed his position in "Quantenmechanik der Stossvorgange" (1926; received July 21, 1926). At that time, he had not yet read the final two papers in the quantization series. He explained the differences between his and Schrödinger's interpretation in a paper, "Physical Aspects of Quantum Mechanics," read at Oxford on August 10, 1926 (after Q3 was published but before Q4). This is reprinted in Max Born, _Physics in My Generation_ (1969, pp. 6-12). An excellent account of the further development of the probability interpretation may be found in Jammer (1966, pp. 281-293; 305-307; 320-321).

[33]Werner Heisenberg, who was working with Bohr during Schrödinger's week-long visit, summarized the Bohr-Schrödinger interaction in "Quantum Theory and Its Interpretation" (1967, pp. 103-104) and in _Physics and Beyond: Encounters and Conversations_ (1971, pp. 73-76).

[34]This letter is microfilmed in the collection of Bohr's scientific correspondence (Note 1, reel 16, sec. 2). The citations translated in the article are: "Hier liegt scheinbar ein Widerspruch vor und Sie sagen: hier

reichen eben unsere bisherigen Worte und Begriffe nicht aus. Ich kann mich bei diesen Konstatierung nicht befriedigen und ich kann daraus nicht für mich das Recht ableiten, mit widerspruchsvollen Aussagen weiter zu operieren. . . . Was mir vor Augen schwebt, ist nur die eine these: man darf, auch wenn hundert Versuche fehlschlagen, die Hoffnung nicht aufgeben, ich sage nicht durch Klassische Bilder, aber durch logisch widerspruchsfreie Vorstellungen von der wirklichen Beschaffenheit des räumlich zeitlichen Geschehens zum Ziele zu kommen. Es is ausserordentlich wahrscheinlich, dass das möglich ist."

[35]Schrödinger (1927a, p. 126) lists the papers that presented independent discoveries of the equation now referred to as the Klein-Gordon equation. Though his results were not published, Pauli's name should be added to the list of Klein, Gordon, de Broglie, V. Foch, H. van Dungen, and T. de Donder. See van der Waerden (1973).

[36]The continuity equation is $\Sigma_1^4 \partial S_\alpha/\partial x_\alpha = 0$, where

$$S_\alpha = 1/\alpha(\bar{\psi}\partial\psi/\partial x_\alpha - \psi\partial\bar{\psi}/\partial x_\alpha - 4\pi i/n(e/c)\Phi_\alpha\psi\bar{\psi})$$

and Φ_α gives the four components of the electromagnetic potential. The first three components of S_α may be interpreted as the spatial distribution of a current density and the fourth component as a charge density.

[37]Dirac's gradual development of a unified theory of quantum mechanics seems to have had a strong influence on Schrödinger. Among his unpublished papers is a 24-page notebook entitled "Dirac" relating Dirac's developments to Schrödinger's own work. This is undated but was probably written in the summer or fall of 1926. A microfilm copy is in the archives for the History of Quantum Theory (Kuhn et al., 1967, folder 41, p. 2).

[38]Heisenberg's (1926a) explanation of the distinction is: "In der physikalischen Interpretation der mathematischen Formalisms besteht jedoch ein Unterscheid zwischen Schrödingers und unserer Auffassung. Wegen der mathematischen Äquivalenz des Schrödingerschen Verfahrens mit den Quantenmechanik könnte allerdings die Frage nach den Gleichungen zugrunde liegenden physikalischen Geschehen eissteilen als eine unsere Anschauung betreffende Zweckmässigkeitsfrage betrachtet werden; aber nur solange

wir nicht versuchen, auf Grund der einmal gewählten anschaulichen Bilder die Grundlage dieser Quantentheories zu erweitern" (p. 411).

[39]See Heisenberg (1926c). This was sent from Copenhagen on July 24, 1926, before Bohr and Heisenberg developed the Copenhagen interpretation.

[40]Schrödinger's paper was received on June 10, 1927; Heisenberg's, on March 23, 1927. However, it was published in an issue of the Zeitschrift marked "Abgeschlossen" on May 31, which I assume to mean 'closed by the editor and sent to the publisher'.

[41]In Heisenberg (1960) and in an interview with T. Kuhn (Kuhn et al., 1967, folder 8).

[42]Other problems, in addition to the disagreement with Schrödinger, contributed to the genesis of this paper. A basic one was Heisenberg's attempt to accommodate the observed paths of electrons in cloud chambers into his theory. See Heisenberg (1971, pp. 77-81).

[43]A similar view had already been presented by Dirac (1927a). This was sent from Copenhagen when Bohr and Heisenberg were still working together on an interpretation of quantum theory and was received in England on December 1, 1926. Heisenberg was familiar with this article (it was cited by Heisenberg, 1927b, on page 176), and with the position presented: "One cannot answer any questions on the quantum theory which refer to numerical values for both the q_{ro} and the p_{ro}" (coordinate and momentum measurements). All that quantum theory can do, Dirac argued, is to give associated ranges of values for canonically conjugate variables. "Questions of this type appear to be the only ones to which the quantum theory can give a definite answer, and they are probably the only ones to which the physicist requires an answer" (Dirac, 1927a, p. 623). Dirac clearly realized the limitations in the information quantum theory could give but did not formulate this in terms of a principle as Heisenberg did.

[44]"Darin erblicken wir den tiefen Sinn der Linearität der Schrödingerschen Gleichungen; deswegen können sie nur als Gleichungen für Wellen im Phasenraum verstanden werden und deswegen möchten wir jeden Versuch, diese Gleichungen z.B. im relativistischen Falle (bei

mehren Electronen) durch nichtlineare zu ersetzen, für aussichtslos halten" (Heisenberg, 1927b, p. 184).

[45]This claim is based primarily on the bibliography given in Scott (1967, pp. 151-161).

[46]Schrödinger's statement (Solvay Conference, 1928) is: "On a trouvé depuis que l'identification naïve d'un électron se mouvant sur une orbite macroscopique avec un paquet d'onde se heurte à des difficultés et ne peut donc pas être admise à la lettre. La difficulté principale est celle-ci que le paquet d'onde s'éparpille certainement de tous côtés quand il rencounte un obstacle, un atom par example. . . . Le compromis proposé de divers côtés, consistant à admettre une association d'ondes et d'électrons ponctuels, je le tiens simplement pour une facon de résoudre la difficulté" (p. 288).

[47]A detailed treatment of this may be found in Herzberg (1950, pp. 82-90). It would be logically possible to follow Schrödinger's interpretation if one also reinterprets rotations and vibrations as nothing but phenomenal manifestations of underlying frequencies. This, however, is a position that few physicists then or now would take seriously.

Note added in proof: I wish to thank Linda Wessels for her scholarly criticism and her insightful presentation of a somewhat different interpretation of the same development. In stressing the competition between Schrödinger and Heisenberg, I did not intend to exclude or minimize the strong concern both men had in developing adequate and acceptable interpretations for the new formalisms they introduced. Creative scientists, however, are not immune to the mixed motives that inspire lesser men. Thus, an agonal striving for competitive superiority characterized Heisenberg's chess, ping pong, mountain climbing, and music, as well as his work in physics. Schrödinger was inclined to compete only on his own terms, but he was certainly a competitor. Unless this neglected aspect is brought out, neither the intensity and timing of Schrödinger's work in 1926 and early 1927 nor the cessation of his creative work after Heisenberg's indeterminacy paper is quite intelligible.

I think that Professor Wessels is quite correct in pointing out that Schrödinger, like Heisenberg, was relieved with the results of his intertranslatability paper. Unless the two formalisms could be so related, there was serious doubt about the adequacy of both of them. My basic point here is that it was this very success that brought the issue of physical interpretation to the forefront. Finally, it must be admitted that the significance I attribute to this competition is a conjecture. Such historical conjectures must be tested and evaluated through historical research. Since Wessels, and not I, is now doing the historical research on Schrödinger's personal correspondence and unpublished manuscripts for this period, she will eventually have the final word.

References

Bohr, N., Kramers, H. A., & Slater, J. C. The quantum theory of radiation. _Philosophical Magazine_, 1924, _47_, 785-802.

Born, M. Zur Quantenmechanik der Stossvorgange. _Zeitschrift für Physik_, 1926, _37_, 863-867. (a)

Born, M. Quantenmechanik der Stossvorgange. _Zeitschrift für Physik_, 1926, _38_, 803-827. (b)

Born, M. Physical aspects of quantum mechanics. In _Physics in my generation_ (2nd rev. ed.). New York: Springer-Verlag, 1969.

Born, M., Heisenberg, W., & Jordan, P. Zur Quantenmechanik II. _Zeitschrift für Physik_, 1926, _35_, 557-615.

Dirac, P. A. M. Relativity quantum mechanics with an application to Compton scattering. _Proceedings of the Royal Society_ (_London_), 1926, _A111_, 405-423.

Dirac, P. A. M. The physical interpretation of the quantum dynamics. _Proceedings of the Royal Society_ (_London_), 1927, _A113_, 621-641. (a)

Dirac, P. A. M. The quantum theory of the emission and absorption of radiation. _Proceedings of the Royal Society_ (_London_), 1927, _A114_, 243-265. (b)

Dirac, P. A. M. The quantum theory of the electron. _Proceedings of the Royal Society_ (_London_), 1928, _A117_, 610-624. (a)

Dirac, P. A. M. The quantum theory of the electron (Part 2). Proceedings of the Royal Society (London), 1928, A118, 351-361. (b)

Dirac, P. A. M. Obituary for Schrödinger. Nature, 1961, 189, 355-356.

Eckart, C. Operator calculus and the solutions of quantum dynamics. Physical Review, 1926, 28, 711-726.

Ehrenfest, P. Bemerkung über die angenaherte Gultigkeit der klassischen Mechanik innerhalb der Quantenmechanik. Zeitschrift für Physik, 1927, 45, 455-457.

Forman, P. Weimar culture, causality, and quantum theory, 1918-1927: Adaption by German physicists and mathematicians to a hostile intellectual environment. Historical Studies in the Physical Sciences, 1971, 3, 1-117.

Gerber, J. Geschichte der Wellenmechanik. Archive for History of Exact Sciences, 1969, 5, 349-413.

Gordon, W. Der Comptoneffekt nach der Schrödingerschen Theorie. Zeitschrift für Physik, 1926, 40, 117-133.

Hanson, N. R. The concept of the positron. Cambridge, England: Cambridge University Press, 1963.

Heisenberg, W. Mehrkörperproblem und Resonanz in der Quantenmechanik. Zeitschrift für Physik, 1926, 38, 411-426. (a)

Heisenberg, W. Quantenmechanik. Naturwissenschaft, 1926, 14, 989-995. (b)

Heisenberg, W. Spektra von Atomsystemen mit zwei Elektronen. Zeitschrift für Physik, 1926, 39, 499-519. (c)

Heisenberg, W. Schwankungserscheinungen und Quantenmechanik. Zeitschrift für Physik, 1927, 40, 501-506. (a)

Heisenberg, W. Über den anschaulichen Inhalt der quantentheoretischen Kinematik und Mechanik. Zeitschrift fur Physik, 1927, 43, 172-198. (b)

Heisenberg, W. Erinnerungen an die Zeit der Entwicklung der Quantenmechanik. In M. Fierz & V. F. Weisskepf (Eds.), Theoretical physics in the twentieth century: A memorial volume to Wolfgang Pauli. New York: Interscience, 1960.

Heisenberg, W. Quantum theory and its interpretation. In S. Rozenthal (Ed.), Niels Bohr. Amsterdam: North-Holland, 1967.

Heisenberg, W. *Physics and beyond: Encounters and conversations* (A. J. Pomerans, trans.). New York: Harper & Row, 1971.

Herzberg, G. *Molecular spectra and molecular structure I. Spectra of diatomic molecules* (2nd ed.). New York: Van Nostrand, 1950.

Hund, F. *The history of quantum theory* (G. Reece, trans.). New York: Harper & Row, 1974.

Jammer, M. *The conceptual development of quantum mechanics*. New York: McGraw-Hill, 1966.

Jordan, P. Über quantenmechaniksche Darstellung van Quantensprungen. *Zeitschrift für Physik*, 1927, *40*, 661-666.

Jordan, P. Early years of quantum mechanics: Some reminiscences. In J. Mehra (Ed.), *The physicist's conception of nature*. Dordrecht: Reidel, 1973.

Klein, M. Einstein and the wave-particle duality. *Natural Philosopher*, 1964, *3*, 3-49.

Kuhn, T. S., Heilbron, J. L., Forman, P., & Allen, L. (Eds.). *Sources for the history of quantum physics: An inventory and report*. Philadelphia: American Philosophical Society, 1967.

Landsberg, G., & Mandelstamm, L. Eine neue Erscheinung bei der Lichtzerstreuung in Krystallen. *Naturwissenschaft*, 1928, *16*, 557-558.

MacKinnon, E. Ontic commitments of quantum mechanics. In *Boston studies in the philosophy of science*, *XIII*. Dordrecht: Reidel, 1974.

MacKinnon, E. A reinterpretation of Harre's Copernican revolution. *Philosophy of Science*, 1975, *42*, 67-79. (a)

MacKinnon, E. Review of Erhard Scheibe's *The logical analysis of quantum mechanics*. *Philosophy of Science*, 1975, *42*, 96-100. (b)

MacKinnon, E. Heisenberg, models and the rise of matrix mechanics. *Historical Studies in the Physical Sciences*, 1977, *8*, 137-188.

MacKinnon, E. De Broglie's thesis: A critical retrospective. *American Journal of Physics*, in press. (a)

MacKinnon, E. Review of Jeffrey Bub's *The interpretation of quantum mechanics*. *Philosophia*, in press. (b)

MacKinnon, E. *Neils Bohr: Philosopher-scientist*. In preparation.

Medicus, H. A. Fifty years of matter waves. *Physics Today*, 1974, *27*, 38-45.

Przibram, K. (Ed.). Letters on wave mechanics (M. Klein, trans.). New York: Philosophical Library, 1967.

Raman, C. V., & Krishnan, K. S. A new type of secondary radiation. Nature, 1928, 121, 501-502.

Salpeter, E. E., Corson, D. R., & Bauer, S. H. Peter J. W. Debye--An interview. Science, 1964, 145, 554-559.

Schrödinger, E. Bohrs neue Strahlungshypothese und der Energiesatz. Naturwissenschaft, 1924, 12, 720-724.

Schrödinger, E. Zur Einsteinschen Gastheorie. Physicalische Zeitschrift, 1926, 27. 95-101. (a)

Schrödinger, E. Quantisierung als Eigenwertproblem (Erste Mitteilung). Annalen der Physik, 1926, 79, 361-376. (Trans. in Schrödinger, 1927a, pp. 1-12.) (b)

Schrödinger, E. Quantisierung als Eigenwertproblem. Zweite Mitteilung. Annalen der Physik, 1926, 79, 489-527. (Trans. in Schrödinger, 1927a, pp. 13-40.) (c)

Schrödinger, E. Quantisierung als Eigenwertproblem, III. Annalen der Physik, 1926, 80, 437-490. (Trans. in Schrödinger, 1927a, pp. 62-101.) (d)

Schrödinger, E. Quantisierung als Eigenwertproblem, IV. Annalen der Physik, 1926, 81, 109-139. (Trans. in Schrödinger, 1927a, pp. 102-123.) (e)

Schrödinger, E. Der steige Übergang von der Mikro--zur Makromechanik. Naturwissenschaft, 1926, 14, 664-666. (Trans. in Schrödinger, 1927a, pp. 41-44.) (f)

Schrödinger, E. Über das Verhaltnis der Heisenberg-Born-Jordanschen Quantenmechanik zu der meinen. Annalen der Physik, 1926, 79. 734-756. (Trans. in Schrödinger, 1927a, pp. 45-61.) (g)

Schrödinger, E. An undulatory theory of the mechanics of atoms and molecules. Physical Review, 1926, 28, 1049-1070. (h)

Schrödinger, E. Collected papers on wave mechanics (J. F. Shearer & W. M. Deans, trans., 2nd ed.). London: Blackie & Sons, 1927. (a)

Schrödinger, E. Energieaustausch nach der Wellenmechanik. Annalen der Physik. 1927, 83, 956-968. (Trans. in Schrödinger, 1927a, pp. 137-146.) (b)

Schrödinger, E. Der Energieimpulssatz der Materiewellen. Annalen der Physik 1927, 82, 265-272. (Trans. in Schrödinger, 1927a, pp. 130-136.) (c)

Schrödinger, E. Über den Comptoneffekt. Annalen der Physik, 1927, 82, 257-264. (Trans. in Schrödinger, 1927a, pp. 124-129.) (d)

Schrödinger, E. *Four lectures on wave mechanics*. London: Blackie & Sons, 1928.

Schrödinger, E. The meaning of wave mechanics. In *Louis de Broglie: Physicien et penseur*. Paris: Albin Michel, 1953.

Schrödinger, E. *Science and the human temperament*. (The unrevised reprint was retitled *Science, theory and man*.) New York: Dover, 1957.

Schrödinger, E. *My view of the world* (C. Hastings, trans.). Cambridge, England: Cambridge University Press, 1964.

Schrödinger, E. Conceptual models in physics and their philosophical value (trans. in *Science, theory and man*, pp. 148-164.)

Scott, W. T. *Erwin Schrödinger: An introduction to his writings*. Amherst: University of Massachusetts Press, 1967.

Smekal, A. Zur Quantentheorie der Dispersion. *Naturwissenschaft*, 1923, *11*, 873-875.

Solvay Conference. *Electrons et photons: Rapports et discussions du Cinquieme Conseil de Physique tenu a Bruxelles du 24 au 29 Octobre 1927 sous les Auspices de l'Institut International de Physique Solvay*. Paris: Gauthier-Villars, 1928.

Tomonaga, S.-I. *Quantum mechanics* (Vol. 2: *New quantum theory*; Koshibå, trans.). New York: Wiley, 1966.

van der Waerden, B. L. *Sources of quantum mechanics*. New York: Dover, 1967.

van der Waerden, B. L. From matrix mechanics and wave mechanics to unified quantum mechanics. In J. Mehra (Ed.), *The physicist's conception of nature*. Dordrecht: Reidel, 1973.

Watson, J. D. *The double helix*. New York: Atheneum Press, 1968.

Wentzel, G. Eine Verallgemeinerung der Quantenbedingungen für die Zwecke der Wellenmechanik. *Zeitschrift für Physik*, 1926, *38*, 518-529.

Wessels, L. *The development of Schrödinger's thought*. Unpublished doctoral dissertation, University of Indiana, 1976.

Reference Note

1. *The Niels Bohr Library*. Microfilm copies of the correspondence in this library are available at various repositories and are referred to by reel number.

THE INTELLECTUAL SOURCES OF SCHRÖDINGER'S INTERPRETATIONS

Linda Wessels
Indiana University

The original controversy over the interpretation of quantum mechanics, dating from the nearly simultaneous discoveries of matrix and wave mechanics in late 1925 to the almost unanimous adoption of the 'Copenhagen interpretation' following the 1927 Solvay Conference, should be of great interest to both historians and philosophers of science. The exchange of arguments and counter-arguments during this period is the process that brought about major changes in most scientists' view of the relation between world and theory. In addition, this controversy reveals how the scientific community reacted to and eventually resolved (at least temporarily) some of the problems of theory comparison currently at issue in philosophical circles, for it arose because that community faced two theories that appeared to have radically different form and physical content, but also had (according to several independently discovered proofs) exactly the same empirical consequences. Clearly, a historical study of this controversy is long overdue.

MacKinnon's account (the first article in this volume) of Schrödinger's attempt to give a 'physical interpretation' to his wave mechanics must be regarded as a welcome beginning to such a study. It was Schrödinger's interpretation that captured the hearts of the still classical-minded majority of the physics community and left Schrödinger's opponents (Heisenberg, Born, Jordan, Dirac, Bohr, and the rest) with the task of trying to change their minds. In the process, the members of this latter group were forced to clarify and eventually to revise and improve the interpretation they advocated. MacKinnon's analysis of Schrödinger's wave mechanics papers reveals both the interpretation around which the controversy centered and some of the problems with that interpretation upon which its opponents fastened. In

Studies in the Foundations of Quantum Mechanics, pp. 59-76

examining the controversy itself, he has concentrated primarily on the exchanges between Schrödinger and Lorentz (in published correspondence) and between Schrödinger and Heisenberg (in some of their articles of 1926-27). His discussion of the latter exchange is particularly interesting, since it highlights the concern over questions of interpretation that lay behind some of Heisenberg's most important early papers--papers now valued primarily for innovations such as the theory of resonant energy exchange, the distinction between symmetric and antisymmetric state functions, and the application of quantum mechanics to explain the multiplet spectral line structure of many-electron atoms and the band spectra of molecules.

To characterize MacKinnon's work as a beginning is certainly not to diminish its value. Much more than a single paper will be required to give a full account of the 1926-27 debates over interpretation, and MacKinnon has chosen to concentrate on a topic central to such an account--Schrödinger's own attempt to give wave mechanics "an adequate and consistent physical interpretation" (p. 1). It is, rather, in his treatment of this restricted topic that I see the need to raise objections. Besides a few (<u>very</u> few) quibbles over historical details which I will not go into here, most of my objections concern the general picture MacKinnon paints of Schrödinger's attitude toward the interpretation problem and motives for trying to develop an electrodynamic interpretation of wave mechanics.

The unbroken thread weaving through MacKinnon's paper is the claim that Schrödinger's battle to find an intuitively satisfying physical interpretation for wave mechanics and to defend this interpretation against all comers was motivated by a fear of losing his priority as the discoverer of <u>the</u> quantum theory. MacKinnon implies that Schrödinger was disappointed when he discovered the intertranslatability of the wave and matrix mechanics formalisms. This seemed to show that the two theories really amounted to the same thing and, thus, to award priority to Heisenberg, who had discovered matrix mechanics almost six months before Schrödinger discovered wave mechanics. MacKinnon goes on to assert that it was this fear that caused Schrödinger to change drastically the focus of his published papers. Before his discovery of intertranslatability, the emphasis was on the empirical adequacy of his interpretationally neutral 'wave' equation; after the discovery, Schrödinger made the question of interpretation a major topic of his papers and worked desperately to come up with and defend an interpretation compatible with the suggestive form of wave mechanics and incompatible with the apparently positivistic views of the architects of matrix mechanics. Finally, MacKinnon outlines the problems that doomed Schrödinger's efforts to failure and speculates that it was a manic low over this failure that induced Schrödinger to drop out of the quantum mechanics game altogether and to move on to other areas of physics.

This thread bears fairly well the weight of the historical data MacKinnon offers. But it will not hold up under the increased load of further evidence drawn from other sources. I will try to show this by unraveling some of the strands of MacKinnon's account and examining them in the light of these additional considerations.

First, consider MacKinnon's central claim--that only after his discovery of the intertranslatability of wave mechanics did interpretation become a "basic rather than a peripheral issue" (p. 11) for Schrödinger. There is an ambiguity here, since each of two readings is compatible with the use MacKinnon makes of this claim. We might understand it as an assertion about a shift in Schrödinger's actual attitude toward the interpretation question or about a change in the sorts of issues that he was willing to emphasize in his published papers. I will argue that the first of these readings renders the claim false, while the second invites damaging criticism of the argument in which the claim appears as a premise. Schrödinger's insistence on an intuitively satisfying physical interpretation for quantum mechanics was just one manifestation of an abiding concern over the physical plausibility of the models underlying the formal structure of scientific theories. This concern shows through any number of his writings both before and after the struggle over interpretation in 1926 and 1927. For example, in 1924 Schrödinger wrote a laudatory article about the recently developed Bohr-Kramers-Slater theory of atomic radiation processes (Bohr, Kramers, & Slater, 1924; Schrödinger, 1924). He expressed his particular pleasure with the basic assumption of that theory--

that the laws governing atomic radiation processes are essentially statistical--since it coincided with his own current views on the nondeterministic nature of microprocesses (Schrödinger, 1929/1957). But in the midst of his praise he voiced one objection and it was aimed at the physical plausibility of the Bohr-Kramers-Slater atomic model. He argued that the atomic electrons cannot be assumed to jump suddenly from one orbit to another during the radiation process, because independent data from various studies on the coherence of the emitted light showed that the resultant wave train was about 10^8 wavelengths long and therefore took about 10^{-8} seconds to produce (Schrödinger, 1924, p. 722). Either Schrödinger did not understand the nondescriptive role that Bohr and his followers had come to assign such models (cf. Bohr, 1928, 1965) or he understood their position but refused to accept it. A 1924 letter to Bohr (Schrödinger, Note 6) shows that the latter view is correct. There Schrödinger objected to the terminology in the Bohr-Kramers-Slater paper that implicitly disclaimed any reality for the waves that coordinated the statistically governed electron transitions. "I cannot really go along with you," he wrote, "when you keep calling these 'virtual'. . . . For what is the 'real' radiation if it is not that which 'causes' transitions?" Schrödinger's opinion of the positivistic dismissal of such questions of reality was recorded a year later, shortly before he launched into the work that eventually led him to the discovery of wave mechanics. Beginning a rather speculative philosophical essay (Schrödinger, 1964), he bade his readers to

> call to mind that sense of misgiving, that cold clutch of dreary emptiness which comes over everybody . . . when they first encounter the description given by Krichoff and Mach of the task of physics (or of science generally): "a description of the facts, with the maximum of completeness and the maximum of economy of thought"; . . . in actual fact (let us examine ourselves honestly and faithfully), to have <u>only</u> this goal before one's eyes would not suffice to keep the work of research going forward in any field whatsoever. (pp. 3-4)

Moreover, an analysis of the steps that led Schrödinger to his famous wave equation shows that wave mechanics was a direct product of his concern over the physical pictures underlying the mathematical superstructure of scientific theories. Schrödinger's route to wave mechanics can be traced from his preparation of a seminar on de Broglie phase waves in early fall of 1925, through an attempt to get a detailed picture of the structure of these waves as they traveled about the atomic nucleus, a subsequent application of the concept of phase waves to ideal gas theory, and finally to a search, inspired by the results of this application, for the wave equation governing the 'electrons' of the atom. In particular, Schrödinger's application of de Broglie's phase wave notion to ideal gases was part of a much larger program to find a physically plausible quantum theory of gases that would give the same gas equations as Bose-Einstein statistics. Such a theory would provide an explanation for the apparent correctness of the new statistics, which were, in Schrödinger's view, merely an ad hoc formal device. The success of Schrödinger's application of the phase wave notion to ideal gases in November of 1925 (1926h) not only demonstrated to him the promise of a wave theory of matter but also forced him to treat these waves as standing waves to be characterized by a wave equation and boundary conditions rather than as traveling waves to be described by construction of a wave function, thus altering his picture of the waves surrounding the atomic nucleus. It was then a quite natural step to begin searching for the wave equation governing the behavior of such standing 'electron' waves. (See Wessels, in press, for a full account.)

Even in his first two papers on wave mechanics (1926a, 1926b), hereafter referred to as Q1 and Q2, one glimpses the wave picture behind the austerity of Schrödinger's interpretationally neutral presentation. In the first of these papers (Q1) he discussed briefly the "more intuitive way" he had "originally intended to found this new version of the quantum conditions," characterizing it as "a vibration process" (1926a, p. 372). A physical mechanism for radiation emission was suggested:

> It is quite conceivable that on the transition of energy from one normal vibration to another, something--I mean the light wave--with a frequency related to each frequency difference makes its appearance. One needs only to imagine that the

> light wave is causally related to the *beats* which are forced to arise at each point of space during the transition, and that the frequency of the light is defined by the number of times per second that the intensity maximum of the beat process repeats itself. (p. 374, italics in original)

Again in his second wave mechanics paper (Q2), even in the midst of the detailed construction of his wave equation on the basis of Hamilton's analogy in which he carefully restricts himself to talk of waves in an abstract multidimensional q-space, Schrödinger slipped in a suggestive description of "a wave group . . . which in some way gets into a small closed 'path'" about an atomic nucleus, and then "not only fills the whole path domain all at once but also stretches far beyond it in all directions" (1926b, pp. 507-508). And, as MacKinnon himself points out, the purely formal requirements put on the solutions to the wave equation so neutrally presented in the first paper clearly reflect the fact that Schrödinger thought of those solutions as related in some way to real waves in physical space.

It is clear that Schrödinger did not view this wave picture as merely a heuristic device--a model useful for theory construction, like Heisenberg's anharmonic oscillators (cf. MacKinnon, 1977), but not to be taken seriously as a description of microprocesses. In Q2 (1926b) Schrödinger stated his opposition to the casual attitude toward physical models that the partisans of matrix mechanics had adopted:

> It has been doubted whether atomic events can even be described in the space-time form of thought. From the philosophical standpoint, I would consider a final decision of this sort to be equivalent to a complete surrender. For we cannot really change our form of thought, and what we cannot comprehend within it we cannot understand at all. (p. 509)

Furthermore, his correspondence of this time shows that he worried a good deal about how his wave interpretation could be filled in and strengthened to give the full space-time picture he thought so important for physics. This is particularly clear in his exchanges with Wien on the radiation emission process (Schrödinger, Notes 8, 11; Wien, 1930, p. 72).

Note that all of the above citations come from sources dating from before Schrödinger's discovery of the intertranslatability of matrix and wave mechanics. Thus, we cannot suppose that this discovery suddenly awakened in Schrödinger a previously nonexistent concern for intuitively satisfying physical interpretations. That concern had been a central part of his attitude toward physical theories prior not only to his discovery of intertranslatability, but even to his discovery of wave mechanics. It was, in fact, the motivating force behind the work on de Broglie phase waves and ideal gases that brought him to the brink of the latter discovery.

But what about the second reading of MacKinnon's claim--that Schrödinger's discovery of the intertranslatability of matrix and wave mechanics brought about a change in the sorts of issues that he was willing to emphasize in his published papers? This certainly cannot be denied. The few bits of a wave picture peeking through Schrödinger's first two wave mechanics papers are overwhelmed by an unmistakable emphasis on the formal aspects of his theory and the technical problems involved in calculating the all-important empirical results. In stark contrast, Schrödinger's fourth paper (1926d), hereafter referred to as Q4, contained his electrodynamic interpretation as a major topic. (In the third paper [1926c], Schrödinger outlined a perturbation theory for wave mechanics and applied it to get the first-order Stark effect, a calculation he had already worked out by the end of February and, thus, before he discovered the intertranslatability; cf. Note 9.) But surely fear of losing priority is not the only reason that can be offered for this shift in emphasis. Let me present an alternative, and I think more plausible, explanation of why Schrödinger began to focus more directly on the issue of interpretation after the intertranslatability of matrix and wave mechanics was proved.

The overriding concern with formal and technical matters in the first two of Schrödinger's wave mechanics papers can be attributed to several factors. First, Schrödinger was presenting a radically new and mathematically sophisticated theory of microphysics to an audience with a variety of preconceptions about what a satisfactory quantum theory should look like. Microphysics was a relatively new field, best known for the intractability of the phenomena it tried to explain and for spawning lots of nice ideas that did not quite work. Many of those in Schrödinger's reading audience were convinced that a clever reworking of classical theories would do the job, some thought that a drastically nonclassical approach was needed, a growing number saw the key in a surrender of any hope for intuitively satisfactory physical pictures (among these were Bohr, Born, Sommerfeld, and their disciples), and still others took such pictures to be the ultimate goal of a scientific theory. The most obvious and direct way to convince such a group that his was a theory worth taking seriously was simply to show that it worked. This was clearly the main purpose of his first three wave mechanics papers. Arguments over interpretation could wait until the correctness of the theory was established. In addition, he had to show how it worked. Wave mechanics included no ad hoc hypotheses (like the old 'whole number requirement') and required no guesswork (like Bohr's 'principle of correspondence'), and, in contrast to both the old quantum theory and the new matrix mechanics, it was simple and elegant in form. But particular applications required some fairly high-powered mathematics (Schrödinger himself had had difficulties in trying to solve the wave equation [cf. Note 7] and needed weekly sessions with Hermann Weyl for help on technical points [Note 5]), and Schrödinger was forced to devote much of the space in his first three papers to outlining calculation procedures and to spelling out the details of the most complicated parts.

Even so, Schrödinger had originally planned to present his wave mechanics in a more intuitive way--presumably by bringing out more clearly the wave picture on which it had been constructed (1926a, p. 372). Certain later developments influenced the writing of Q1 and Q2, however, in a way that focused these papers on more formal aspects of the theory. Erwin Fues (Notes 1, 2), one of Schrödinger's young assistants at Zurich, has indicated that it was a suggestion from Peter Debye that shaped the Q1 presentation. Fues recalls that prior to publishing Q1, Schrödinger gave two seminars on his theory for his Zurich colleagues. In the second of these seminars (presumably held in early January), Debye remarked that all of the most fundamental physical theories can be founded on extremum principles; he advised Schrödinger to fashion the same sort of elegant formulation for wave mechanics. Schrödinger must have turned to this task immediately, for by the end of January he had submitted Q1, in which the wave equation is derived as the result of an extremum requirement on a version of the classical Hamilton-Jacobi equation. The derivation was given no intuitive motivation (indeed, it has none); coupled with the subsequent detailed demonstration of how to solve the wave equation for the hydrogen atom, it naturally placed the emphasis of Q1 on the mathematical properties of Schrödinger's theory.

The Q2 development of the wave equation, in which one considers purely abstract 'waves' in a multidimensional coordinate space, was also a freshly conceived presentation. We do not know how Schrödinger came upon the idea of building on Hamilton's analogy to get wave mechanics, but it clearly happened only after he had already written Q1 (Wessels, in press). He was apparently so enthusiastic about this new way of developing the wave equation that he again set aside his original plans for an intuitive approach and centered his next paper on this more abstract presentation. (Schrödinger also went on to use this new approach by way of Hamilton's analogy in his later discussions of wave mechanics--see, e.g., Schrödinger, 1926g, pp. 1050-1056; 1928a, p. ix; 1928b, pp. 1-9; 1933/1957.)

The ease with which Schrödinger was sidetracked from his original plan for an intuitive presentation was probably due to the fact that when he began writing his series of papers on wave mechanics, there were still difficult problems facing his intended interpretation. Analysis of his papers and correspondence is enough to give a pretty good idea of what that interpretation was and of the difficulties that concerned Schrödinger the most (cf. Wessels, 1975, chap. 3). The problem of wave packet dispersion, which MacKinnon pinpoints, was not perceived as insurmountable but rather as a difficulty that would eventually be worked out. Thus, in Q2 he conjectured that a way could be found to construct nonspreading wave-mechanical packets (Schrödinger, 1926b, p. 505), and right after he finished Q2 he took time out from his wave mechanics series to demonstrate that it could in fact be done for the quantum-mechanical harmonic oscillator (Schrödinger, 1926e). He still had hope for the packets as late as June 6 (several months after his discovery of intertranslatability, while in the midst of working out Q4), when he sent a copy of this paper to Lorentz in answer to the latter's query about wave packet dispersion. Schrödinger expressed hope in the accompanying letter that similar constructions could be carried out for at least 'quasi-periodic' motions--though he seemed ready to admit that "the electron is incapable of existing in a completely field-free space . . . perhaps even that 'free' electrons do not permanently keep their identities" (Przibram, 1965, p. 59).

There was an apparently intractable problem however, arising from the fact that the solutions to the wave equation for a compound system (a system consisting of two or more interacting subsystems) were functions of more than three spatial coordinates. Thus, if these solutions described waves at all, they were waves in an abstract n-dimensional coordinate space. The only hope of reconciling this with the wave picture that had been suggested by Schrödinger's work on quantum gas theory--and that formed the basis for his original, intended interpretation of wave mechanics--was to recast the wave equation for compound systems in a form that yielded three-dimensional wave functions as solutions. This hope was very dim indeed. It is no wonder that in the presence of this and other problems (cf. Wessels, 1975, pp. 212-213) Schrödinger chose to sidestep the interpretation issue until he had ironed out a few more wrinkles.

Schrödinger's discovery of the intertranslatability of matrix and wave mechanics brought about a more explicit concern with the interpretation issue in his papers, not because it raised the priority question, but because it yielded as one of its by-products a new, more promising physical interpretation. The origins of matrix mechanics suggested certain relations between that theory and the electromagnetic theory of radiation; the formal translation of matrix to wave mechanics allowed this suggestion to be transferred to wave mechanics, where the quite natural result was Schrödinger's electrodynamic interpretation (cf. Wessels, 1975, pp. 240-243). As MacKinnon points out, Schrödinger's first version of this interpretation, in which $\psi \frac{\delta\psi^*}{\delta t}$ was associated with electric charge density, had obvious problems--the imaginary part of the product had to be arbitrarily discarded, the integral of the product over all space is zero, and for a compound system the product gives a density only in an abstract configuration space. All of these problems were avoided when Schrödinger adopted his second version of this interpretation, however, in which $\frac{\delta\psi^*}{\delta t}$ was replaced by ψ^*, and a special procedure for calculating the charge density for compound systems was spelled out. (Apparently Schrödinger first used the partial derivative of ψ^* because he thought it would make the application of the interpretation to his relativistic version of wave mechanics more convenient--cf. Schrödinger, 1926f, p. 476.) The first version of this interpretation was suggested tentatively in the paper containing Schrödinger's proof of intertranslatability; the second version was introduced and described in some detail three months later, in Q4. At that time, Schrödinger saw no major difficulties with the interpretation, apparently expecting to spend the next few months working out a few loose ends and thus complete the task he had begun only eight months before (cf. Wessels, 1975, pp. 267-272). It was probably only several weeks after the completion of Q4 that he learned of Born's radically different interpretation of wave mechanics, and perhaps several months before he became aware of some genuine problems facing his own interpretation and began to sense that in defending his views he was facing an uphill struggle.

According to this alternative explanation, then, the shift in emphasis that MacKinnon points out stems not from fear about priority but from Schrödinger's new confidence in what he had to say about interpretation. The issue of interpretation had been an important one for him all along, but the old wave picture originating in his work on quantum gas theory had become inadequate once his new wave mechanics had taken form and it was not until his discovery of the intertranslatability of matrix and wave mechanics that he saw the clue to a defensible interpretation--one that still gave an appropriately intuitive picture and seemed to pose no obvious problems. I think that the plausibility of this alternative account will be strengthened as we examine several more strands of MacKinnon's account.

In presenting his account, MacKinnon conjectures that Schrödinger was disappointed when he found that matrix and wave mechanics were intertranslatable. It seems to me, however, that just the opposite was the case. Once he had completed several applications of his theory, Schrödinger recognized that his and Heisenberg's theories were in remarkable agreement when it came to empirical predictions--particularly striking was the fact that both theories gave the same unexpected energy spectrum for the one-dimensional oscillator (Schrödinger, 1926b, p. 516). Several passages in Q2 indicate that Schrödinger had already been thinking about other similarities between the two theories and expected that eventually some clearly definable relation between the two would be discovered. In a February 22 letter to Wien, Schrödinger (Note 8) explicitly expressed his conviction that there was "an inner connection" between his and Heisenberg's theory. But, he explained, the connection "apparently lies deep," since even Weyl, whom he had consulted on this question, could not find it. Though Schrödinger concluded by telling Wien that he had therefore given up his own efforts to find "the connecting link," it was less than one month later that he sent his paper on intertranslatability (1926f) to Wien for publication and exclaimed in the accompanying letter: "The hope of which I spoke . . . in my February 22 letter has been fulfilled much more quickly than I thought. The relation to Heisenberg's work is now completely clear, and indeed in the sense that whoever does not want to calculate using matrices need not, because the two representations are completely equivalent from a purely mathematical standpoint" (Schrödinger, Note 10, italics added).

Schrödinger's reaction to this discovery seems to have been one of relief--as was Heisenberg's. The two theories were no longer competitors at the level of empirical consequences. And each adopted the position that he could now appropriate the computational advantages of the other's theory (in the case of matrix mechanics, the ability to determine radiation intensities; in the case of wave mechanics, the much more familiar and elegant apparatus of partial differential equations), while simply discarding the attendant interpretation as spurious. Thus, Schrödinger obtained wave mechanical expressions for line intensities via the translation from matrix to wave mechanics and went on to construct his electrodynamic interpretation around these expressions. Heisenberg reports that prior to the discovery of intertranslatability, he had deliberately ignored Schrödinger's theory because of the interpretation attached to it. Only after he saw that the two theories were formally equivalent did he take a serious look at wave mechanics; he then quickly began using the wave mechanical formalism, although simply as an auxiliary calculating device for getting matrix mechanical results (Heisenberg, Note 3). It was Bohr who took seriously the fact that the intertranslatability made it possible to give two drastically different interpretations to either form of quantum mechanics. Right from the beginning he insisted that the correct interpretation (which still remained to be formulated) would somehow have to incorporate both the discontinuous quantum picture underlying matrix mechanics and the continuous oscillatory picture underlying wave mechanics (Heisenberg, Note 3; Hund, Note 4).

Of course, both Heisenberg and Schrödinger saw the immediate implication of their positions--that the major issue now separating them was that of interpretation. And it was with that issue that they and the rest of the physics community wrestled for more than a year. MacKinnon's supposition that Schrödinger's part in that struggle was motivated only (or even primarily) by an emotional concern over his priority is surely too simple. Emotions did indeed run high during the controversy under interpretation. But if we take the word of the participants in that controversy, the emotions arose because of strong convictions about the correctness of their own positions. Thus Schrödinger's initial joy at having found "a picture in which at least something or other really takes place with that frequency which we observe in the emitted light," his outburst over "all this damned quantum jumping" during his visit to Copenhagen, his refusal to give up "hope of reaching the goal of a representation of the true properties of space-time events," and his stubbornly persistent efforts to defend and reformulate his interpretations in the face of mounting evidence against them are more plausibly seen as manifestations of his Boltzmannian belief in the necessity of finding intuitively satisfactory physical pictures than as desperate attempts to rationalize a losing battle for priority. We have already seen that

Schrödinger did not suddenly acquire this belief when the intertranslatability threatened his claim to historical immortality, but that it also shaped his pre-wave-mechanics attitude toward theories of microprocesses. And Schrödinger's later papers (e.g., 1947/1956a, pp. 183, 191, 196; 1947/1956b, pp. 26, 237; 1950/1957, pp. 203-204; 1958, p. 169) show that this belief remained a part of his conception of science long after his place in the history of science had been secured.

Thus, for Schrödinger, the crisis precipitated by the discovery of intertranslatability was not so much a personal one, centered on fears of personal gain or loss, but rather an intellectual one, threatening some of his basic convictions about the nature of science. For the discovery of intertranslatability gave rise to a long and difficult controversy over interpretation that involved most of the major physicists of the day; it was the attendant process of argument and counterargument and eventual resolution that in turn forced Schrödinger to question his insistence on the necessity of physical pictures. At first, Schrödinger thought he had the upper hand. His new electrodynamic interpretation looked promising, a number of prominent physicists were in sympathy with his approach, and the partisans of matrix mechanics did not have a very clear alternative to offer. But in the course of the controversy, Schrödinger gradually became aware of the problems facing his interpretation, even discovering some of them himself (Notes 12, 13, 14). None of these problems constituted a refutation in itself, but their increasing number and variety had to be accepted as weighty evidence against the plausibility of his approach. I suspect that it was the pressure of the consolidating opinion of his professional colleagues that finally forced him to face the failure of his interpretation and to abandon his attempts to generate alternatives to the newly accepted 'Copenhagen interpretation'. Schrödinger reacted to this crisis by retaining his long-held convictions about the necessity of intuitive pictures and all but giving up his active participation in the further development of quantum theory.

The alternative account that I have sketched of Schrödinger's motives for constructing and defending his intuitive interpretations of wave mechanics is not without its own weak spots. Some readers will argue that it is as one-sided in its concern with intellectual influences as MacKinnon's is with psychological causes, and that the correct account is surely some combination of the two. That may be. My purpose has not been to argue that Schrödinger was a cold, purely rational scientist, without ego or uncontrolled emotions. This was most certainly not the case. It has been, rather, to show that there is strong evidence for the claims that underlying Schrödinger's position in the original controversy over interpretation was a concern for substantive issues, that when he seriously backed an interpretation of wave mechanics it was because he had good reason to think it was a correct interpretation, and that his emotion-charged defense of that interpretation was grounded in his firm convictions about the need for such interpretations in a satisfactory scientific theory.

I have aimed my arguments in this direction not because I think that psychological analysis has no place in a historical study of science but because in this case the evidence indicates that intellectual considerations were a prime mover. Because Schrödinger and his supporters eventually turned out to be wrong, it is easy to look back and think that if only they had considered their position coolheadedly they would have seen their mistakes immediately--that their resistance to the truth must be chalked up to pure stubbornness. But, in the initial stages of the controversy over interpretation, _nobody_ had a clearly articulated position, let alone a handle on the 'truth'. The discovery of the intertranslatability of wave and matrix mechanics posed some difficult problems of theory comparison and interpretation. The incompatible efforts to solve these problems forced physicists to face seriously the problem of interpretation, to formulate their positions more carefully, and to come up with solid arguments in defense of these views and against those in opposition. The 'Copenhagen solution' finally accepted by most physicists shortly after the 1927 Solvay conference was not a contender in the struggle but the ultimate product of this dialectic process. Viewed thus, as a struggle over substantive issues rather than as a priority battle, the controversy promises to reveal much about argument and conceptual change in science and about the sources of the current traditions in quantum theory.

References

Bohr, N. Uber die Anwendung der Quantentheorie auf den Atombau. Zeitschrift für Physik, 1928, 13, 117-165.

Bohr, N. The structure of the atom. In Nobel lectures, Physics II, 1922-1941. Amsterdam: Elsevier, 1965.

Bohr, N., Kramers, H. A., & Slater, J. C. The quantum theory of radiation. Philosophical Magazine, 1924, 47, 785-802.

MacKinnon, E. Heisenberg, models, and the rise of matrix mechanics. Historical Studies in the Physical Sciences, 1977, 8, 137-188.

Przibram, K. (Ed.). Letters on wave mechanics (M. J. Klein, trans.). New York: Philosophical Library, 1965.

Schrödinger, E. Bohrs neue Strahlungshypothese und der Energiesatz. Die Naturwissenschaften, 1924, 12, 720-724.

Schrödinger, E. Quantisierung als Eigenwertproblem (Erste Mitteilung). Annalen der Physik, 1926, 79, 361-376. (a)

Schrödinger, E. Quantisierung als Eigenwertproblem (Zweite Mitteilung). Annalen der Physik, 1926, 79, 489-527. (b)

Schrödinger, E. Quantisierung als Eigenwertproblem (Dritte Mitteilung). Annalen der Physik, 1926, 80, 437-490. (c)

Schrödinger, E. Quantisierung als Eigenwertproblem (Vierte Mitteilung). Annalen der Physik, 1926, 81, 109-139. (d)

Schrödinger, E. Der stetige Ubergang von der Mikro- zur Makromechanik. Die Naturwissenschaften, 1926, 14, 664-666. (e)

Schrödinger, E. Uber der Verhältnis der Heisenberg-Born-Jordenschen Quantenmechanik in der Meinen. Annalen der Physik, 1926, 79, 734-756. (f)

Schrödinger, E. An undulatory theory of the mechanics of atoms and molecules. Physical Review, 1926, 28, 1049-1070. (g)

Schrödinger, E. Zur Einsteinschen Gastheorie. Physikalische Zeitschrift, 1926, 27, 95-101. (h)

Schrödinger, E. Collected papers on wave mechanics J. F. Shearer & W. M. Dean, trans.). London: Blackie & Son, 1928. (Originally published: Abhandlungen zur Wellenmechanik (2nd ed.). Leipzig: Barth, 1928.) (a)

Schrödinger, E. Four lectures on wave mechanics. London: Blackie & Son, 1928. (b)

Schrödinger, E. On the peculiarity of the scientific world-view. In What is life? and other scientific essays. Garden City, N.Y.: Doubleday Anchor, 1956. (Originally published: Acta Physica Austriaca, 1947, 1, 201-245.) (a)

Schrödinger, E. The spirit of science. In What is life? and other scientific essays. Garden City, N.Y.: Doubleday Anchor, 1956. (Originally published: Der Geist der Naturwissenschaft. In Eranos Jahrbuch. Zurich: Rhein, 1947.) (b)

Schrödinger, E. What is a law of nature? In Science, theory and man. New York: Dover, 1957. (Originally published: Was ist ein Naturgesetz? Die Naturwissenschaften, 1929, 17, 9-11. Schrödinger's inaugural address for the University of Zurich, December 9, 1922.)

Schrödinger, E. The fundamental idea of wave mechanics. In Science, theory and man. New York: Dover, 1957. (Schrödinger's 1933 Nobel Prize Address.)

Schrödinger, E. What is an elementary particle? In Science, theory and man. New York: Dover, 1957. (Originally published: Endeavour, 1950, 9, 109-116.)

Schrödinger, E. Might perhaps energy be a merely statistical concept? Nuovo Cimento, 1958, 9, 162-170.

Schrödinger, E. Seek for the road. In My view of the world. Cambridge, England: Cambridge University Press, 1964.

Wessels, L. Schrödinger's interpretation of wave mechanics. Doctoral dissertation, Indiana University, 1975.

Wessels, L. Schrödinger's route to wave mechanics. Studies in the History and Philosophy of Science, in press.

Wien, W. Letter to E. Schrödinger, March 22, 1926. In Wilhelm Wien: Aus dem Leben und Wirken eines Physikers. Leipzig: Barth, 1930.

Reference Notes

Unpublished materials listed below are part of the Archive for History of Quantum Physics (below abbreviated as AHQP). See Sources for History of Quantum Physics: An Inventory and Report, edited by T. S. Kuhn, J. L. Heilbron, P. Forman, and L. Allen (published by the American Philosophical Society, Philadelphia, 1967), for a catalogue of its contents and a listing of its locations.

1. Fues, E. Letter to T. S. Kuhn, July 2, 1963, AHQP (Reel 66, Section 8).
2. Fues, E. Letter to T. S. Kuhn, August 1, 1963, AHQP (Reel 66, Section 8).
3. Heisenberg, W. Interview with T. S. Kuhn, February 25, 1963, AHQP.
4. Hund, F. Interview with T. S. Kuhn, June 25, 1963, AHQP.
5. Schrödinger, A. Interview with T. S. Kuhn, April 5, 1963, AHQP.
6. Schrödinger, E. Letter to N. Bohr, May 24, 1924, AHQP (Bohr Scientific Correspondence Reel 16).
7. Schrödinger, E. Letter to W. Wien, December 27, 1925, AHQP.
8. Schrödinger, E. Letter to W. Wien, February 22, 1926, AHQP.
9. Schrödinger, E. Letter to M. Planck, February 26, 1926, AHQP (Reel 41).
10. Schrödinger, E. Letter to W. Wien, March 19, 1926, AHQP.
11. Schrödinger, E. Letter to W. Wien, May 6, 1926, AHQP.
12. Schrödinger, E. Letter to G. Joos, November 17, 1926, AHQP (Reel 41).
13. Schrödinger, E. Letter to H. A. Kramers, November 19, 1926, AHQP (Reel 41).
14. Schrödinger, E. Letter to N. Bohr, November 25, 1926, AHQP (Bohr Scientific Correspondence Reel 16).

AN OPERATIONAL ANALYSIS OF THE DOUBLE SLIT EXPERIMENT

H. Pierre Noyes
Stanford Linear Accelerator Center

I. INTRODUCTION

This paper illustrates the contention that quantum mechanical effects can be reduced to statistical distributions of yes-no events corresponding to counts in detectors. We claim that these distributions correspond to objectively distinguishable classes of events, and that their probabilities add without interference. In our view, noncommuting 'observables' and complementarity are confined to the model space employed by the quantum mechanical formalism in the calculation of these probability distributions; our 'objective reality' resides in the individual events, the macroscopic measurements which specify their approximate spatial and temporal relations, and the finite distributions which result from any particular test of the theory. All other aspects of the problem are discussed within the model space, and are posed under the restriction that the model space allows us to carry out the implied operations.

We hold that classical physics also employs a model space. This space differs subtly from the model space of quantum mechanics in that it allows limiting points of space and velocity (or time) to be defined independently in terms of abstract 'particles'. In contrast, we will discover in our specific context that the model space of quantum mechanics, also using abstract 'particles', easily allows the velocity limit to be taken (given enough space) but can define short distances only indirectly. Hence the space-time points in the model space do not have the same operational significance as those in the model space of classical physics, even though both are treated formally as mathematical continua.

We believe that one major source of confusion in discussions about the foundations of quantum mechanics is

Studies in the Foundations of Quantum Mechanics, pp. 77-108

the often made but usually unstated assumption that the model space of classical physics is identical to `real physical space'--whatever that means. This confusion becomes worse once it is also assumed that `particles' are `real physical objects' whose positions and velocities can be defined in that space. For us this last assumption is patently a metaphysical assertion rather than a scientific statement. Once that assertion is made, or even worse, assumed without explicit recognition of the fact that this has been done, the fact that quantum mechanics employs a model space with different properties becomes distorted into a belief that quantum mechanics requires the `real world' to have paradoxical, alogical, or even antilogical properties. All these problems can be avoided by not falling into that trap in the first place.

It has been forcefully pointed out to us that by confining the essential discussion to the model spaces of classical or quantum physics we disengage just where the philosophical problems become most interesting. Our excuse is that most of the problems we avoid are common to classical and quantum physics, at least from our own point of view. We are not willing to grant that physicists exist in the abstract, let alone the "Great Mathematician" of Jeans' cosmology. For us, either classical or quantum physics has the burden of showing that it is capable of a scientific retrodiction of the past which leads to physicists, using "historical" data compatible with the model in question; a brief effort of ours along these lines has been attempted (Noyes, Note 1). Most of this discussion does not involve explicitly the points at issue between classical and quantum physics. So far as we can see, any evolutionary or dialectical scheme which leads from physical cosmology through terrestrial biological development to the cultural evolution of physicists must, in current scientific context, use both conserved elements (matter in the broadest sense) and statistical elements on which selection can act in a unidirectional manner leading to increased complexity. We have argued elsewhere (Noyes, Note 2) that the absolute conservation laws for discrete quantum numbers coupled with the fixed-past uncertain-future interpretation of quantum mechanics (Noyes, 1975) form a more satisfactory scientific basis for such schemes than does classical continuum physics, but obviously cannot pursue this further here. In what follows we assume that, to requisite accuracy, physicists can measure "real" macroscopic space and time intervals, and agree whether counters fire or not whatever their basic beliefs about the models they correlate with these measurements.

The paradigm we choose to illustrate our point of view is the double slit experiment. In order to bring out the objective nature of the predictions for this experiment, we employ particles of finite mass rather than 'light quanta' and assume that the two slits enclose particle detectors which sometimes detect the passage of a particle and sometimes do not. We then predict that three different statistical distributions will be detected in an array of particle detectors placed a sufficient distance behind the two slits. The usual double slit interference pattern corresponds to the case when neither of the slit detectors fires, while the two other patterns arise when one or the other of the slit detectors fires in addition to one of the detectors in the final array. These three distributions are predicted to add without interference, and to be objectively distinguishable by making use of appropriate time gates on the detectors and (delayed) coincidences between the detector signals.

Corrections to this initial prediction arise from nonuniform illumination of the slits and can be calculated from a detailed discussion of the way in which the beam of particles striking the slits is prepared. Additional corrections arise from the detection events in the detectors themselves and can be calculated from a model of the elementary process that activates the detector. We show that these corrections can, in principle, be made arbitrarily small, thus justifying the simpler treatment with which we start. We also discover that we do not actually have to perform coincidence measurements to distinguish the three distributions, because these distributions change in predictable ways as we change the density of the material in the detectors. This allows us to extrapolate our macroscopic thought experiment down to atomic dimensions and relate it to actual measurements of the interference effects observed between de Broglie waves. Some known limitations on this extrapolation are discussed in our concluding section.

II. FIRST APPROXIMATION TO THE PROBABILITY DISTRIBUTIONS

The model of the double slit experiment we consider is presented in Figure 1. We assume that particles of mass M and momentum close to some value p_0 (velocity $V_0 = p_0/M$) traveling in the z direction are produced by an accelerator or some other source. For simplicity we assume uniformity in the dimension perpendicular to the figure until we come to discuss corrections. To select particles with a known uncertainty in momentum we employ a collimator consisting of two slits (also containing particle detectors) of width W a distance S apart; the exit detector D_C is in a slit a distance C from the double slit arrangement we are studying. If we pick our zero of time as the time when a particle from the source is supposed (classically) to pass through one of the two slits, the momentum of this particle is ensured to be close to p_0 by gating the detector D_S closest to the source to be active at time $t \sim -(S + C)M/p_0$ and the detector D_C to be active at time $t \sim -CM/p_0$. We collect data only for cases in which D_S and D_C both fire while these time gates are open. Our third requirement is that the final detector array D_3 fire at time $t \sim +rM/p_0$, where $r = (y^2 + z^2)^{1/2}$ is the distance from the center of the double slit arrangement to the coordinates y,z describing the center of the detector in the array that fires. In addition to these three requirements (i.e., D_S, D_C, and D_3 must fire during the time intervals Δt set by the time gates centered around the two times given above) we also record whether or not D_1 or D_2 fires during a time interval Δt_0 centered on $t = 0$. We assume that the intensity is low enough so that corrections due to pileup (e.g., D_1 and D_2 both firing during this time interval) are not severe.

The prediction of the quantum mechanical model is that we can distinguish three different distributions in the detector array D_3. These are defined as: (I_0),

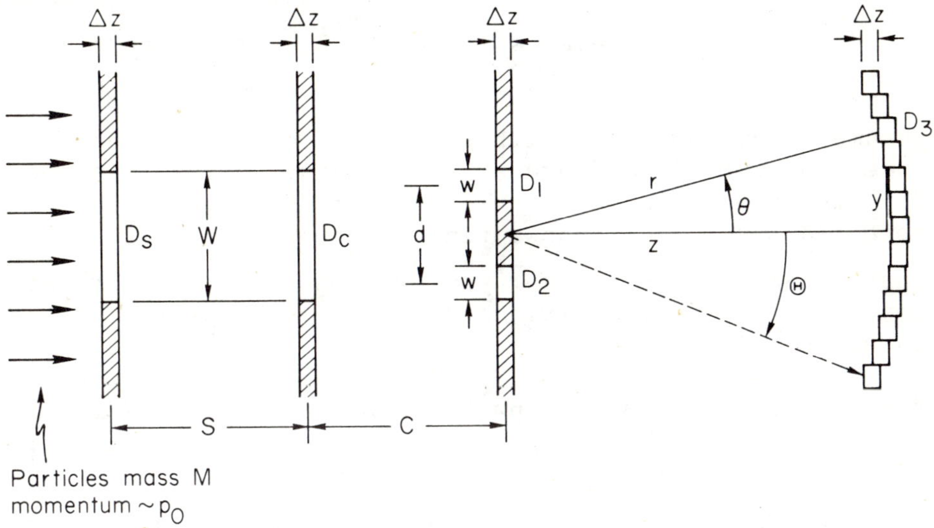

$$D_S \text{ fires at } t = -(S+C)\,M/p_0 \pm \frac{\Delta t}{2}$$

$$D_C \text{ fires at } t = -cM/p_0 \pm \frac{\Delta t}{2}$$

$$(I_0): D_3 \text{ fires at } t = rM/p_0 \pm \frac{\Delta t_3}{2}$$

$$(I_1): D_1 \text{ fires at } t = 0 \pm \frac{\Delta t_0}{2} \quad D_3 \text{ fires at } t = rM/p_0 \pm \frac{\Delta t_3}{2}$$

$$(I_2): D_2 \text{ fires at } t = 0 \pm \frac{\Delta t_0}{2} \quad D_3 \text{ fires at } t = rM/p_0 \pm \frac{\Delta t_3}{2}$$

4-77 3155A1

FIGURE 1. Center section of the geometrical arrangement for the experiment. The minimal dimension of the detectors is Δz, so w, d, and W are larger. Dimensions of the detectors in the x direction (perpendicular to the plane of the figure) are ℓ for the detectors in the slits and L for the detectors in the counters. The minimal time resolution Δt exceeds $M\Delta z/p_0$ (see text). If Δt_0 and Δt_3 are less than $\Delta t_{min} = M\Delta z/p_0$, events which would be allowed in the data set as valid are excluded. Optimal values for Δt_0 and Δt_3 depend on the intensity (i.e., number per unit time of valid events), since high intensity leads to an increase in the number of valid events, but also to a more rapid increase in the number of ambiguous events. The maximum value of Δt_3 which will ensure the maximum value of valid events allowed by the collimator is given in the text, but the optimization of this conjointly with the intensity is not performed.

those events in which neither D_1 nor D_2 fires when $t \sim 0$; (I_1), those events in which D_1 fires when $t \sim 0$; and (I_2), those events in which D_2 fires when $t \sim 0$. We assume that both the slit widths w and the distance between the two slits d are small compared to the distance to the detector array r. We further assume that the detector array D_3 subtends a finite angular aperture $\pm\Theta$, that the angles $\theta = \tan^{-1} y/z$ within this aperture are small compared to d/r, and that $r \gg \hbar/p_0$. We define three functions

$$f_S^{\pm}(\theta;w,k/p,d/r,\Theta) = \sin^2[pw(\theta \mp d/2r)/\hbar]/N_S(\theta \mp d/2r)^2 \tag{1}$$

$$f_d(\theta;w,d,\hbar/p,\Theta) = \sin^2[pw\theta/2\hbar]\cos^2[pd\theta/2\hbar]/N_d\theta^2 \tag{2}$$

with

$$N_S = \int_{-\Theta}^{\Theta} d\theta f_S(\theta) \; ; \qquad N_d = \int_{-\Theta}^{\Theta} d\theta f_d(\theta) \; . \tag{3}$$

Then the prediction for these three distributions, assuming a total of N data, is

$$I_0 = N(1 - e_1 - e_2) f_d(\theta;w,\hbar/p_0,d/r,\Theta) \tag{4}$$

$$I_1 = Ne_1 f_S^{+}(\theta;w,d,\hbar/p_0,\Theta) \tag{5}$$

$$I_2 = Ne_2 f_S^{-}(\theta;w,d,\hbar/p_0,\Theta) \; . \tag{6}$$

These are, of course, the usual double and single slit diffraction patterns for wavelength $\hbar/p_0$ in the small angle approximation. The parameters e_1 and e_2 are simply the fraction of the total number of data for which D_1 or D_2 fires. The prediction is statistical; for large N it will be approached in the sense of the law of large numbers. The fractions e_1 and e_2 are predicted to be proportional to the detector efficiencies, as will be discussed in Section IV.

We note that this single experimental arrangement suffices to provide data exhibiting both single slit and double slit interference patterns. This appears to conflict with loose statements about complementarity, such as "an experiment which allows us to determine which slit the particle passes through is complementary to an experiment which yields the double slit pattern." We see that if by experiment we mean a single event, it is correct to say that the three possibilities belong to disjoint and objectively distinguishable classes of events and hence in that restricted sense are complementary. However, if the term 'experiment' means, as it usually does in physics, a system capable of producing numbers which can be compared with the theoretical prediction, the statement is palpably false.

In order to test the prediction, we must accumulate enough data to compare the prediction, at some confidence level, with the large number limit. If it happens that e_1 and e_2 are very small, we will get a reasonable check on the prediction of I_0 for a smaller value of N than we need to check the predictions for I_1 and I_2; but if $e_1 + e_2$ is close to unity, the reverse will be true. Since e_1 and e_2 are adjustable by changing the density of the detecting systems in the detectors, we can obviously choose conditions such that our confidence in the test of all three predictions reaches roughly the same level for the same value of N. Further, for the same reason, if we collect data for various values of e_1 and e_2 we can verify the prediction without even making use of the time tagging of the three distributions. Hence, allowed this freedom, the experiment can be 'overdetermined'; this is important from an experimental point of view, since it allows us a check on some types of systematic error.

We now derive the approximate results quoted above. Concentrating first on states of a single momentum p traveling in the +z direction, if there were no boundaries, the wave function would be $\psi(z,t) = \exp i(pz/\hbar - p^2t/2M\hbar)$. The corresponding flux is $j_s = (\hbar/2Mi)(\psi^*\partial\psi/\partial z - \psi\partial\psi^*/\partial z) = p/M = v$ particles per unit 'area' per unit time; in our two-dimensional problem, 'area' is unit length perpendicular to the z direction, while in three dimensions it is area in the xy plane. We assume that our source and collimator are such that, to a first approximation, our two slits are illuminated with a plane wave of this type and that the detector does not significantly alter the momentum of the individual particles in this beam. We also assume that the randomness in phase allowed by the first two time gates is uniform enough so that only the time-average value of ψ matters. All of these approximations will be justified subsequently. They allow us to look for a solution of the time-independent Schrödinger equation $(\partial^2/\partial z^2 + \partial^2/\partial y^2 + p^2/\hbar^2)\psi(y,z) = 0$ in the half space $0 \leq z < \infty$, $-\infty < y < \infty$ which vanishes on the line $z = 0$ except for two uniform distributions of equal magnitude for y lying in the two slit intervals $(d + w)/2 > y > (d - w)/2$ and $-(d - w)/2 > y > -(d + w)/2$.

Since we also need the single slit pattern, we first consider a single slit of width w centered at $y = 0$, and then show how the desired result can be derived from that. Our boundary condition for this simpler problem is that $\psi_w(y,0) = \frac{1}{w}[\theta(y + w/2) - \theta(y - w/2)]$. The solution must be of the form

$$\psi_w(y,z) = \int_{-\pi/2}^{\pi/2} d\phi f_w(\phi) e^{ip \cos \phi z/\hbar} e^{ip \sin \phi y/\hbar} . \tag{7}$$

Hence, by taking the Fourier transform of the boundary condition we find that

$$f_w(\phi) = (1/\pi w)\sin(pw \sin \phi/2\hbar)\cot \phi \tag{8}$$

or, letting $\xi = \sin \phi$, that the single slit wave function is

$$\psi_w(y,z) = \frac{1}{\pi w}\int_{-1}^{1} \frac{d\xi}{\xi} \sin(pw\xi/2\hbar) e^{ipz\sqrt{1-\xi^2}/\hbar} e^{ipy\xi/\hbar} \tag{9}$$

or, in polar coordinates,

$$\psi_w(r \sin \theta, r \cos \theta) = \frac{1}{\pi w} d \frac{\sin(pw\xi/2\hbar)}{\xi} \cdot e^{i(pr/\hbar)(\sqrt{1-\xi^2} \cos \theta + \xi \sin \theta)} \tag{10}$$

If we had used a rectangular aperture of finite length ℓ in the x direction as well as finite width w, we could clearly have obtained a similar result using two angles for the resolution of p in the x, y, and z directions, and ended up with a somewhat more complicated double integral representing $\psi_{w\ell}$.

In order to obtain the desired result we now make use of our assumption that the detector array D_3 is so far away from the slits that $pr/\hbar >> 1$. We can then evaluate the integral by the method of stationary phase, as explained, for example, by Jeffries and Jeffries (1950), obtaining

$$\psi_w \to \left[\frac{2\pi\hbar(1 - \xi_0)^2}{pr}\right]^{1/2} \frac{\sin(pw\xi_0/2\hbar)}{w\xi_0} e^{i(pr/\hbar - \pi/4)} + 0(\hbar/pr) \tag{11}$$

where

$$\xi_0 = \sin \theta \; . \tag{12}$$

If detector 1 fires, we know that the particle passed through the upper slit. We can therefore obtain the wave function for this case by using the method just developed with our boundary condition displaced upward by d/2, rather than being centered at y = 0. The result is again Equation 11 with $\xi_0 \to \xi_1$, defined by

$$(\xi_1)^2 = \frac{(\sin \theta - d/2r)^2}{\cos^2\theta + (\sin \theta - d/2r)^2} \; . \tag{13}$$

If detector 2 fires the result is obviously given in the same way with $\xi_0 \to \xi_2$ and

$$(\xi_2)^2 = \frac{(\sin \theta + d/2r)^2}{\cos^2\theta + (\sin \theta + d/2r)^2} \; . \tag{14}$$

If neither 1 nor 2 fires, our boundary condition spreads the same intensity over the two slits, and the two contributions add coherently, yielding the wave function

$$\psi_{2w,d}(r\cos\theta, r\sin\theta)$$

$$= \frac{1}{\pi}\int_{-1}^{1} d\,\frac{\sin(pw\xi/2\hbar)}{w\xi}\cos(pd\xi/2\hbar) \cdot e^{i(pr/\hbar)(\sqrt{1-\xi^2}\cos\theta + \xi\sin\theta)} . \tag{15}$$

For this wave function the stationary phase is given by $\xi = \sin\theta$. Using these three wave functions to calculate the flux through D_3 and neglecting θ^2 compared to θ, we finally obtain the three results given in Equations 1-3, as was to be proved.

III. CORRECTIONS DUE TO THE STRUCTURE OF THE COLLIMATOR AND THE SOURCE

This section and the next deal with what is customarily called `preparation of the initial state´ and `measurement´. Elementary treatments simply assume, as we have done in the last section, that it is possible to start with a beam of particles with precisely defined momentum $\underline{p}_0$, and that the measurement corresponds to the detection events in D_3 assumed proportional to $\psi^*\psi$ integrated over the space-time volume defined by that detector and its time gate. I made these assumptions myself in my own derivation of the scattering formalism starting from quantum mechanical particles and particle detectors (Noyes, 1976). Yet this definition of measurement as proportional to $\psi^*\psi$ is hard to relate to the abstract formulation of quantum mechanics as given by von Neumann, as will be readily appreciated after reading the discussion of this problem given by Nancy Cartwright in another article prepared for this volume. The assumption that the initial momentum is well defined is known to be suspect because of the uncertainty principle. Qualitatively we can argue that with all past time available to prepare the beam, both its energy and direction can be defined as sharply as we wish. More precise treatment, for example, that by Goldberger and Watson (1964), estimates the size of the wave packet allowed by the uncertainty principle, and shows that this introduces no problems for the particle beams in actual use, but implicitly assumes that some function $f(p - p_0)$ used to define the wave packet centered on p_0 can itself be known to arbitrarily high precision. We have gone to considerable trouble to show that this assumption can indeed be justified using only particle detectors to measure momentum distributions.

Because of the complexity of the argument, I have found it useful to break my analysis into three parts. First, I assume that the geometrical configuration of the two detectors in the collimators and the time gates applied to them define the momenta which emerge and, under the assumption that the detection events are uniformly distributed over these space-time volumes, calculate the probability distribution of the momenta of the particles thus selected. Then I ask under what restrictions the quantum mechanical model allows us to treat this calculation as a first approximation, with bounded corrections that can be made arbitrarily small, still assuming that the detection events themselves within these geometrical and temporal limits can be treated as punctiform. Finally, I discuss assumptions which suffice to allow the quantum mechanical detection events to be approximated in this manner.

Returning to Figure 1 and the statement of the conditions for data collection, we see that the firing of the two detectors D_S and D_C guarantees (except for the accidental coincidences that will occur if the flux of particles is too intense) that the probability of the particle being in the two space-time volumes

$$-\frac{L}{2} < x_S < \frac{L}{2}, \qquad -\frac{W}{2} < y_S < \frac{W}{2},$$

$$-S - C - \frac{\Delta z}{2} < z_S < -S - C + \frac{\Delta z}{2}$$

$$-\frac{M(S + C)}{p_0} - \frac{\Delta t}{2} < t_S < -\frac{M(S + C)}{p_0} + \frac{\Delta t}{2} \qquad (16)$$

and

$$-\frac{L}{2} < x_C < \frac{L}{2}\,, \qquad -\frac{W}{2} < y_C < \frac{W}{2}\,,$$

$$-C - \frac{\Delta z}{2} < z_C < -C + \frac{\Delta z}{2}$$

$$-\frac{MC}{p_0} - \frac{\Delta t}{2} < t_C < -\frac{MC}{p_0} + \frac{\Delta t}{2} \tag{17}$$

is unity; this implies that the baffles around the slits and the interior walls of the volumes containing the detectors are 'perfectly absorbing', or in other words that any particle that fails to meet these boundary conditions does not reach the two slits under study <u>or</u> return to either detector in the collimator. Corrections due to the failure of this approximation constantly harass the experimental physicist but do not concern us in this thought experiment.

For simplicity we will ignore the lateral (x,y) dimensions and calculate the probability distribution in $p_z = M(z_C - z_S)/(t_C - t_S)$ implied by these boundary conditions. We further assume, to begin with, that the probability of finding a particle within the specified ranges of time and distance is uniform. The quantity of physical interest, the probability distribution in the momentum p_z, depends only on the intervals $z_C - z_S$, and $t_C - t_S$. This fact allows us to define two new variables $z = z_C - z_S - S$ and $t = t_C - t_S - MS/p_0$, and integrate over the uniform distributions in t_S and z_S to obtain

$$dP(z,t) = \Theta\left(\frac{z}{\Delta z}\right)\Theta\left(\frac{t}{\Delta t}\right)\left(1 - \frac{|z|}{\Delta z}\right)\left(1 - \frac{|t|}{\Delta t}\right)\frac{dzdt}{\Delta z\Delta t} \tag{18}$$

where

$$\Theta\left(\frac{x}{\Delta x}\right) \equiv \theta\left(\frac{x}{\Delta x} + 1\right) - \theta\left(\frac{x}{\Delta x} - 1\right)\,. \tag{19}$$

The final step is then to change variables to p_z and some second variable which can be integrated out. In order to bring out more clearly the physical magnitudes involved, we introduce the dimensionless parameters and variables

$$\varepsilon = p_0 \Delta t / MS, \qquad f = \Delta z / \varepsilon S,$$
$$q = (p_z - p_0)/\varepsilon p_0, \quad \xi = \Delta t / t \ . \tag{20}$$

The time a particle of velocity p_0/M takes to cross a length Δz is $M\Delta z/p_0$. If the time resolution Δt were less than this value, we could use the timing device to reduce the spatial resolution Δz. We therefore assume that $\Delta t \geq M\Delta z/p_0$, which implies that f is less than or equal to unity. The momentum p_0 and the mass M are specified in advance. As we will see in the next section, the thickness of the detector Δz depends on what we are willing to assume about its construction. However, the length of the collimator S is still at our disposal, which allows us to make ε as small as we like if we are willing to go to a large enough apparatus for preparing the beam. For similar obvious geometrical reasons, the lateral spread in momentum is proportional to W/S and L/S, and these two dimensions (if bigger than the limiting dimension Δz) are also at our disposal. We can therefore guarantee that the momentum p is well defined in direction as well as in magnitude, if there are no additional wave mechanical restrictions.

Making use of these definitions, we can change variables from z,t to q,ξ and using

$$\frac{z}{\Delta z} = \frac{1}{f}\,[q(1 + \varepsilon\xi) - \xi] \tag{21}$$

find that the probability distribution in q is

$$\begin{aligned} dP(q) &= \frac{dq}{f}\int_{-1}^{1} d\xi(1 + \varepsilon\xi)(1 - |\xi|)(\xi)(1 - |\tfrac{1}{f}\,[q(1 + \varepsilon\xi) - \xi|) \\ &\quad \cdot \left(\tfrac{1}{f}\,[q(1 + \varepsilon\xi) - \xi]\right) \qquad (22) \\ &= [\theta(q + \frac{1 + f}{1 + \varepsilon}) - \theta(q - \frac{1 + f}{1 - \varepsilon})]F(q)dq \ . \end{aligned}$$

For $f \approx 1$ (optimum time resolution) and small ε

$$F(q) \approx \frac{2}{3} - \frac{3}{2}|q| + 2q^2 - |q|^3 \qquad |q| < 1$$
$$\approx \frac{1}{6}(2 - |q|)^3 \qquad 1 < |q| < 2 . \tag{23}$$

The net effect of the calculation is to show that the collimator does indeed produce a momentum distribution peaked about p_0 with the limits

$$p_0 - \varepsilon p_0 \frac{1 + f}{1 + \varepsilon} < p_z < p_0 + \varepsilon p_0 \frac{1 + f}{1 - \varepsilon} . \tag{24}$$

If the source illuminating the collimator does not provide an absolutely uniform distribution of momenta p_z in this range, the assumption that the spatial and temporal distributions within the counters are uniform is not justified due to the correlations between them introduced by this fact. However, the classical calculation is easy to correct if we know the source distribution $P_S(p_z)$, the obvious result being $P_S(p_z)F((p_z - p_0)/\varepsilon p_0)dp_z/\bar{P}$, where $\bar{P}$ is the integral of $P_S F$ over the interval in Equation 24. Since we can construct collimators of arbitrarily high quality by making them long enough, we can measure the probability distribution P_S to any requisite accuracy. This is part of what any experimental physicist would do to calibrate the equipment used in the experiment and so is usually assumed without question. The difficulty from the point of view of this paper is that the calibration so far described uses classical physics and hence cannot be invoked by us if we aim at a self-consistent operational analysis. Even careful treatments such as that of Goldberger and Watson (1964) content themselves with the assumption that the source provides a wave packet $(P_S(p_z))^{1/2} = f_S(p_z)$ whose size can be estimated from the size of the source and with showing that quantum mechanical corrections due to 'wave packet spreading' are, under current conditions, unimportant until the apparatus extends from the earth to the moon or farther. We try to provide below a somewhat more careful analysis. However, to do this we find that it is necessary to invoke some background material which, though familiar to physicists, cannot be expected to form part of the background philosophers and mathematicians invariably bring to the study of the foundations of quantum mechanics.

In classical wave motion, the maximum amplitude A of a monochromatic wave $A \sin [(z - vt)/\lambda\!\!\!^- + \delta]$ and the phase δ are physically measurable quantities. For instance, for a vibrating string, A is the maximum displacement of the string from its equilibrium configuration at position z along the string, and δ can be determined, if the motion is slow enough, by measuring the displacement as a function of time relative to some clock. However, for sound waves in which the amplitude is the average displacement of the air molecules from their equilibrium position, or for light waves in which it is the maximum strength of the electromagnetic field vector, neither A nor δ is experimentally accessible from a practical point of view. Usually for any pitch, the ear, and for all colors, the eye, respond not to the displacement (whose average value is obviously zero) but to the <u>intensity</u> I, which is the square of the displacement averaged over one cycle; clearly, $I = 1/2\ A^2$. Therefore, it is convenient to represent the wave by a complex function $I^{1/2} \exp i[(z - vt)/\lambda\!\!\!^- + \delta] = \psi$, whose absolute square $\psi^*\psi = I$ performs the averaging automatically. Clearly, if we measure only the intensity of a <u>single</u> wave, the phase δ remains unobservable.

At this point, one might well ask how the wave character of either sound or light can be demonstrated. For instance, the source of light might be a candle flame containing $\sim 10^{18}$ individually excited molecules emitting wave trains randomly phased with respect to each other. If we bring beams from two different candles together, there is no interference; the intensities simply add. But if we use a single candle to illuminate a single slit (through a collimator or lens, or far enough away so that the candle can be geometrically approximated by a point source) and look on a far screen, the path length through different points across the width of the slit to the point on the screen can differ by a wavelength; we then can observe the single slit diffraction pattern due to the interference between elements of a <u>single</u> wave train coming from a single atomic source. If the sources are effectively at a point, the <u>intensities</u> on the screen add to give the visible pattern predicted for each separate wave train. The calculation is equivalent to the derivation we gave in Section II. If we use two slits illuminated by a <u>single</u> source, we can observe the double slit pattern. We see that the quantum mechanical calculation for a single particle is like the classical calculation for a single wave train of unknown phase. All that our detectors D_S, D_C, and D_0 do in the particulate case is to ensure, by our time tagging, that we are dealing with a single wave train with the further simplification, if $\Delta z/S$ is small, of a well-defined wavelength. However, if either D_1 or D_2 fires in addition, we destroy any possibility of phase coherence between the two slits, and obtain only the appropriate single slit pattern. The quantum mechanical character of the calculation is thus restricted to the discrete (yes-no) character of the individual particulate detection events, as claimed in our introduction.

The point I hope to have brought out by this digression is that for quantum mechanics the absolute phase of the wave is <u>in principle</u> unknowable, in contrast to the classical situation where the 'random phase approximation' is simply a practical method for dealing with physical situations in which a well-defined quantity happens to be beyond the reach of direct experimental measurement. In my previous discussion of the scattering process (Noyes, 1976), I made use of this principle to derive the conventional scattering formalism by noting that if the absolute coordinates of the positions where scattering takes place entered the theory, they would have to enter the theory as 'quantum mechanical hidden variables' unless the sum over all such processes is performed in a particular way. This enabled me to separate the kinematic (i.e., descriptive) aspect of scattering theory from the dynamical aspect (i.e., the calculation of scattering amplitudes, including relative but not absolute phases). But in that treatment I still assumed that an incident beam of precisely defined momentum could be used as a starting point, an assumption which the discussion presented here is intended to justify.

As we will discuss in more detail in the next section, the detector can be so designed (under restrictions which turn out to be conceptually important) that the detection event does not in itself significantly perturb the incident wave's momentum. Thus if we know $f_S(p_z)$, but not its phase, we can assume that the scattering event that initiates the detection in D_S results in a wave function, for $t > t_S$, which can be written as

$$\psi_C(z,t) = \int_{-\infty}^{\infty} dp e^{i\delta(p)} f_S(p) e^{i[(z-z_S)p/\hbar-(t-t_S)p^2/2M\hbar]} . \quad (25)$$

Under our assumption that this process is incoherent with respect to the source, the phase δ can have any value, and, as just discussed, the parameters z_S and t_S can be known only to the extent that they lie within the ranges given by Equation 16. To justify this assumption a fortiori, we must be able to show that by making this posit the collimator can be used to measure the probability distribution $P_S(p) = f_S^2(p)$ independent of any knowledge of $\delta(p)$; clearly we are assuming both f_S and δ are real and that f_S is appropriately bounded. As in the classical calculation, we assume that all values of z_S and t_S are equally probable within the specified ranges.

If we are correct in assuming that this is the wave function that results from the first detection event, we can then use it to calculate the probability of the second detector (D_C) firing during the gated time interval as proportional to $\psi_C{}^*\psi_C$ when z and t are in the ranges specified by Equation 17. Hence,

$$dP(x_S,t_S,x_C,t_C) \propto dx_S dt_S dx_C dz_C \cdot \int_{-\infty}^{\infty} dp \int_{-\infty}^{\infty} dk f_S(p) f_S(k) e^{i(\delta(p)-\delta(k))} \cdot e^{i[(z_C-z_S)(p-k)/\hbar-(t_C-t_S)(p^2-k^2)^2/2M\hbar]} . \quad (26)$$

As in the classical calculation, we can immediately integrate over t_S and z_S to obtain

$$dp(z,t) = \frac{dz\,dt}{\Delta z \Delta t}\left(1 - \frac{|z|}{\Delta z}\right)\left(1 - \frac{|t|}{\Delta t}\right)\Theta\left(\frac{z}{\Delta z}\right)\Theta\left(\frac{t}{\Delta t}\right)$$

$$\cdot \int_{-\infty}^{\infty} \delta p \int_{-\infty}^{\infty} dk\ f_S(f) f_S(k)$$

$$\cdot\ e^{i(\delta(p)-\delta(k))} e^{i\delta(p-k)\left[1 - \frac{(p+k)}{2p_0}\right]/\hbar} \tag{27}$$

$$\cdot\ e^{i[z(p-k)/\hbar - t(p^2-k^2)/2M\hbar]}.$$

But this result can be evaluated by the method of stationary phase, and we recover immediately the classical result provided only that $S \gg \hbar/p_0$ or, in words, that the distance between the two counters in the collimator is a large number of de Broglie wavelengths. Thus, provided the detectors themselves act as point detectors, when integrated over the active volumes and time gates, we have given a rigorous justification of the classical approximation and have proved that we can measure momentum in a quantum mechanical system to arbitrarily high precision.

The extension of the discussion to three dimensions is straightforward and contains no new conceptual points so far as the time dependence goes, so we will not attempt it here. With regard to the lateral dimensions, physicists will recognize that, in order to avoid diffraction effects, the dimensions W and L will also have to be large when measured in de Broglie wavelengths. In a thought experiment we can always ensure this while still keeping W/S and L/S small simply by scaling everything up in all three dimensions; in any case, the calculation of the _intensity_ of the diffraction patterns across the faces of the counters is a straightforward problem in physical optics, for which standard methods can be used. These can then be used to weight the integrals needed in the probability calculation. Optimization of the counting rates so as to reduce ambiguities due to pileup, now that we have reduced the problem to a classical one, is again a familiar experimental task and provides no conceptual problems. All that remains is to show that, given an initial momentum distribution whose magnitude is knowable to arbitrarily high accuracy, we can indeed construct detectors that will fit into the slits of width w whose diffraction patterns (single and double) we are trying to measure.

IV. CORRECTIONS DUE TO THE STRUCTURE OF THE DETECTOR

The detector we assume to complete our design consists of particles of mass $m \ll M$, struck by the massive particles under study and recoiling. If the recoil energy $p^2/2m$ is greater than some critical energy ε_0, the recoil particle can activate a secondary process, which is amplified by the detector and eventually leads to the signal that we have referred to so far as the firing of the detector. In practice, the energy ε_0 could be the ionization energy of the atoms in the detector, and the amplification process could be an electron and ion cascade, as in a Geiger tube; alternatively, the ionization could produce a photon when the atom that has been ionized recombines, and the amplification could be provided by a photosensitive surface and a photomultiplier tube. But for our purposes all we need is the kinematics of the initiating event, the threshold energy, and the cross section πa_0^2 which each initial event has with respect to the incident beam.

Given this model for the detector, our first step is to calculate the kinematics of the initial scattering process. Assuming that the particles m are free, we can calculate the angular dependence of the momentum and energy of the (scattered) beam and (recoil) detector particles from the conservation of momentum and energy by solving the three equations

$$p \sin \phi = P \sin \theta$$

$$p \cos \phi + P \cos \theta = P_0 \tag{28}$$

$$\frac{p^2}{2m} + \frac{P^2}{2M} = \frac{P_0^2}{2M}$$

with the result

$$P_{(\theta)} = \frac{P_0 \cos\theta + \sqrt{\frac{m^2}{M^2} - \sin^2\theta}}{1 + m/M} \tag{29}$$

$$p^2_{(\theta)} = \frac{2mP_0^2\left[\sin^2\theta + \frac{m}{M} - \cos\theta\sqrt{\frac{m^2}{M^2} - \sin^2\theta}\right]}{M(1 + m/M)^2}.$$

We see immediately that the maximum angle of scattering of the beam particles is $\theta_x = \sin^{-1} m/M$ and hence that simply by using detector particles of small enough mass we can guarantee that the detector does not introduce significant corrections into our analysis so far as angular distributions go. Since the final detector array has angular aperture Θ, all we need require is that $m/M \ll \Theta$.

If the density of detector particles in the volume of the detector V is ρ grams/cm^3, the number of particles is $\rho V/m$, and the cross-sectional area they present to the beam is $\pi\rho V a_0^2/m$. Since the volume of the detector is $w\ell\Delta z$, and the area presented to the beam is $w\ell$, the fraction of the time one beam particle incident on this area will activate the counter is $\pi\rho a_0^2\Delta z/m$. If $\rho = \rho_1$, the density of detector particles in D_1, this fraction is simply e_1, i.e., the fraction of the time that, on the average, the detector will fire when (as our procedure for data collection ensures) exposed to one particle passing through the system. Clearly we must make the density low enough so that $e_1 + e_2 < 1$ for the experiment to work as designed, which also ensures that the corrections due to the beam striking more than one particle within the detector will be small. Thus our detector design ensures the conditions assumed in our initial discussion, and justifies the assertion that we can get equivalent results simply by using particles of known mass m, density ρ, and cross section a_0 in the slits without actually recording whether or not individual data points in the final distribution correspond one by one to the three distributions predicted.

A number of corrections are needed before we can directly confront theory with experiment even under these restrictions. The kinematical equations show us that the momentum of the detector particles varies from 0 when the particle of mass M continues to move in the forward direction ($\sin\theta = 0$) to $p_x = (m/M)[2m/M(1 + m/M)]p_0^{1/2}$ when $\sin\theta = m/M = \sin\theta_x$. In order to activate the detector at all, we must clearly require that $(mp_0/M)^2/M(1 + m/M) > \varepsilon_0$, but there will also be a critical angle θ_C, defined by

$$p_0\left[\sin^2\theta_C + \frac{m^2}{M^2} - \cos\theta_C\sqrt{\frac{m^2}{M^2} - \sin^2\theta_C}\,\right] = M(1 + \frac{m}{M})\varepsilon_0 , \tag{30}$$

within which the detector will not be activated. If the differential cross section for scattering is $\sigma(\theta)$, with $\pi a_0^2 = \int_0^{\theta_x}\sigma(0)d\theta$, then the fraction of the time the detector will not be activated even though scattering occurs is

$$f(m, a_0) = \int_0^{\theta_C}\sigma(\theta)d\theta/\pi a_0^2 . \tag{31}$$

But, as we saw in our last section, the scattering destroys the phase coherence necessary for the double slit interference pattern even though the scattering is not directly observed (i.e., does not activate the detector). Consequently, we must modify our initial prediction to read

$$I_0 = N[(1 - e_1 - e_2)f_d + f(m,\theta_0)(e_1 f_w^+ + e_2 f_w^-)]$$
$$I_1 = Ne_1(1 - f(m,a_0))f_w^+ \tag{32}$$
$$I_2 = Ne_2(1 - f(m,a_0))f_w^- .$$

This makes the comparison with experiment slightly more complicated in practice but does not change anything in principle, except that the correction to I_0 blurs the double slit pattern by including a single slit contamination from those particles which actually scattered from a detector particle in one of the two slits but were not demonstrated to do so because of this detection inefficiency.

In using this expression, we should note that the fractions e_1 and e_2 which occur are the <u>theoretical</u> expressions

$$e_1 = \pi\rho_1 a_0^2 \Delta z/m , \qquad e_2 = \pi\rho_2 a_0^2 \Delta z/m \tag{33}$$

for any scattering event and not the experimental fractions actually measured in any particular experimental run. The statistical prediction for the experimental fractions (i.e., the number of times the detectors D_1 and D_2 fire compared to the total number of events in which D_S, D_C, and D_3 fire) is obviously

$$e_1^{exp} = e_1(1 - f(m,a_0)), \quad e_2^{exp} = e_2(1 - f(m,a_0)) . \tag{34}$$

Comparison between these predictions and the experimental results therefore gives us an internal check on the consistency of our assumptions about the detectors; failure of this check would point to systematic error in the overall setup and suggest that we make independent measurements of ρ_i, a_0, $f(m,a_0)$, the momentum distribution emerging from the collimator or arriving at the detector array D_3, etc., in order to resolve them. Since we have already demonstrated that we can, in principle, construct collimators that allow us to measure momenta to arbitrarily high precision in both magnitude and direction, these experimental checks can all be carried out. They are part of the procedure in any carefully designed experiment, and will not concern us further here.

So far we have not discussed the length of the time gate Δt_3 during which we need to keep the counter D_3 open. We saw in the last section that the collimator does not provide a precisely monochromatic (single momentum) beam but introduces a spread in momentum given by Equation 24. This means that the slowest particles will arrive at D_3 still farther behind the center of the pulse than they leave the collimator and the fastest particles still farther ahead of it. This is a classical 'wave packet spreading' produced by our finite spatial and temporal resolutions and has nothing to do with the quantum mechanical effect of the same name. So long as our time gate is centered on $t = rM/p_0$ we will catch the central portion of the pulse, so using the time gate that will catch <u>all</u> particles that have passed our collimation conditions and the two slits can be simply a matter of optimization of design to increase the counting rate of useful events. However, since this time is necessarily longer than Δt, and hence determines the probability that an event we assign to a particular $D_S + D_C$ trigger is in fact correctly assigned, we need this number to optimize the overall intensity of the incident beam. This number is

$$\Delta t_3 = \left(\frac{S + C + r}{S + \Delta z} \right)\left(\Delta t + \frac{M\Delta z}{p_0}\right) . \tag{35}$$

For similar reasons we should set the time gate Δt_0 on the detectors D_1 and D_2 to be long enough to accommodate the full pulse defined by the collimator. If we make it shorter, we can also make Δt_3 shorter; this suggests that we could simplify the overall design by omitting the counter D_C and using the time gate on the array D_3 to define the momentum. We have not done this, because we thought it simpler conceptually to separate the action of the collimator from the rest of the experiment, rather than present an integrated and optimized design. Another correction, due to the fact that the detection process (even for $\theta < \theta_C$) robs the beam particles of energy and hence slows them down compared to the particles in the double slit pattern, should also be included for completeness. However, this correction, which in principle would require us to use an angle-dependent time gate on D_3 correlated with the firing of D_1 or D_2, and with a corresponding modification of $f(m,a_0)$, would take us into the problem of detecting the time structure in the counts at D_3. It also would give us still another experimental handle on the discrimination between the three patterns I_0, I_1, and I_2; but since we started only with the problem of measuring intensity distributions, we will not pursue the analysis further.

Our treatment so far is not explicitly wave mechanical. We now show that, provided the range R of forces between the beam and detector particles is small compared to the dimensions of the detector, the same approximations suffice. We introduce the relative coordinate $\underline{x} = \underline{r}_M - \underline{r}_m$ between the detector and beam particles, with conjugate momentum $\underline{k} = (mP - Mp)/(M + m)$ and center of mass coordinate $\underline{X} = (M\underline{r}_M + m\underline{r}_m)/(M + m)$ conjugate to the total momentum $\underline{K} = \underline{P} + \underline{p}$. Then the wave function outside the range of forces (i.e., for $x > R$) is simply

$$\psi(\underline{\bar{x}},\underline{x}) = e^{i\underline{K}\cdot\underline{\bar{X}}/\hbar}\left[e^{i\underline{k}_0\cdot\underline{x}/\hbar} + \frac{f(\hat{k}_0\cdot\hat{\underline{x}})e^{ikx/\hbar}}{x}\right]. \tag{36}$$

In the usual treatments of scattering we are interested in the second term, which we examine outside the region to which the initial beam is geometrically confined. In fact, this is how we would measure the cross section for the detection process, since $\pi a_0^2 = (\pi/2)\int_{-1}^{\cos\theta_C} d\xi|f(\xi)|^2$.

In order for our detector not to fire every time a particle goes through we must require that $\pi a_0^2 < w\ell$ (even if there is only one particle in the detector) and $1/n$ of that value if there are n particles. But our geometrical conditions already require us to have $w \lesssim d \ll r$, so that by the time the scattered wave reaches D_3 the scattered amplitude is down by $\sim a_0/r \sim d/r$

(and the intensity by $[a_0/r]^2$) compared to the leading term, which is just the initial free particle wave function (times the detector particle wave function). Further, simply for geometrical reasons, the range of forces R within which Equation 36 has to be modified must be much smaller than w in order to fit the detecting systems into the slit without introducing significant wall effects. We conclude that, if $m/M << \Theta$, $a_0 < w$, and $R << w$, the detector will function as postulated in Section II and that the calculation of corrections, though complicated in detail, can be carried out in terms of experimentally accessible quantities. Further corrections due to nonuniform distribution of the density ρ of detector particles across the slits, momentum dependence of the cross section, nonuniformity of the amplification process which starts from the recoil particle of mass m and ultimately determines in practice the time-resolution characteristics of the detector, not to mention other woes that will occur to an experimental physicist, can also be included. They enter the calculation in the integration of the probability of detection over the detector volumes and the time gates. Since we have already discussed in the last section how these integrals are to be calculated from a quantum mechanical wave function, assuming uniform distributions, we do not repeat the discussion here using these experimental resolution functions; they are simple multiplicative factors of the integrand from a mathematical point of view.

V. DISCUSSION AND CONCLUSIONS

We have now completed our operational analysis of how the double slit experiment can be divided into three objectively distinguishable classes of events, how these are related to counts in detectors with finite spatial and temporal resolutions, and to a limited extent how the corrections to the simple prediction can be either measured or bounded. Our analysis proceeded by two steps. In the first, we showed that we can, in fact, measure momentum to arbitrarily high precision by an appropriate collimator design and hence make the beam incident on the slits as 'monochromatic' as we wish. We then showed that, provided only that the events which initiate the firing of the detectors are single scatterings from particles of mass m much smaller than the mass M of the beam particles under study, the cross section for these scatterings πa_0^2 is much smaller than the cross-sectional area of the detectors in the slits and that the range of force R between the beam particles and the detector particles is small compared to the smallest limiting geometrical dimension (usually Δz, the thickness of the detectors in the beam direction), we can also make the perturbation due to the detectors as small as we wish.

This last restriction on the operation of the experiment, which requires us, in principle, to be able to imagine the existence of particles with arbitrarily small masses and dimensions, is, we believe, the most interesting result to emerge from our analysis. For a strict operationalist (Bridgman, 1928), the concept of length is not the same at atomic dimensions as for lengths that can be measured by laying down a standard ruler a certain number of times. Our attitude is somewhat different. We are exploring the properties of a model space by performing thought experiments and endeavoring to discover what physical model systems are needed to give operational meaning to the theory in question. We reserve discussion of whether this model space is connectable to physical phenomena until this discussion of the model space is completed. Thus, our preliminary conclusion is that, under the restrictions given above, the double slit experiment can be given an objective, operational description. The reason this is possible is that nonrelativistic quantum mechanics, as a theory, makes no assertions about the values of the masses, the cross sections, the range of forces, or the geometrical dimensions that are allowed; the only universal constant that enters is $\hbar$, and, if we concentrate on a particular type of particle, M. In other words, the structure of the theory is completely independent of the units we use for mass, length, and time; nonrelativistic quantum mechanics is <u>scale invariant</u>. This is not so obvious in terms of historical development, since the theory developed out of study of the Bohr model for the atom using first electrostatic and later electrodynamic forces as an integral part of the theory. You will note that I have carried through the above discussion without invoking any electromagnetic effects, partly for reasons of simplicity, but also just to make this point.

To put the matter more graphically, for the experiment we have described, we could, given world enough and time, perform the experiment with cannonballs, the slits being holes in armor plate, the surfaces covered with tar so as to make them 'absorbing'. For a detector we could shoot a stream of buckshot across the openings for a finite time (the 'time gates' Δt), with a detection corresponding to one of the buckshot being knocked out of the stream and either detected separately or found missing by weighing or counting the ones that arrive on the other side of the slit. Someone may object that with cannonballs (at 18th-century speeds) we could simply see which slit the ball went through and thus force the paradox which some see by finding that even so there is (or is not) a double slit interference pattern. But this does not work. If there is sufficient illumination to see each cannonball as it passes through one or the other slit, the radiation pressure on the cannonball is sufficient to destroy the phase coherence needed for the interference pattern to appear. We could use an optical detector, but only at such low light intensity that there is a finite probability that not a single quantum will be absorbed by the cannonball throughout its entire trip through the apparatus. In other words, we are constrained by the theory to perform the experiment in the dark!

Turning now to physics, the least massive particle known is (and has been for nearly a century during which hosts of other 'particles' have been discovered) the electron. Allowing for the moment the possibility of much less massive particles, our conceptual analysis is still adequate, and we can ask what size apparatus we would need to measure, say, the single and double slit diffraction patterns of two slits 0.1 cm wide and 5 cm apart, using electrons in the lowest easily accessible experimental range of velocities, which gives a de Broglie wavelength of about 10^{-8} cm. In order to give sufficient intensity to the double slit pattern, we ask that the first minimum of the single slit pattern from slit 1 fall at the position of the maximum from slit 2. This requires us to choose $r \approx \pi pwd/2\hbar \approx 78.5$ km! Thus, unless we go to supercooled systems (which would allow us to an order of magnitude shorter wavelength) and microscopic slits, we could hardly expect to raise the funds for a direct experimental test of macroscopic de Broglie wave interference. Microdesign and low velocity pose additional problems, since electrostatic image forces induced in the slits by the charge of the electrons, whether the slits are made of conducting or insulating material, would probably override the quantum effect we are seeking. Even so, we could not perform the experiment as described above, since we lack detecting particles smaller than the electron.

The actual tests of the de Broglie wavelength hypothesis are therefore much less less direct than the experiment we have been discussing. They have to assume that the particles under study can be used as 'detectors' by calculating the probabilities of interaction using the wave functions as functions of space and time coordinates (unmeasurable in operational terms) defined by electromagnetic forces. As Bridgman pointed out, so long as only electromagnetic forces are available, it is not clear that these space-time coordinates that occur in the theory have an 'objective' significance, since the operations that define them can only be performed in one way. Fortunately for us, Colella, Overhauser, and Werner (1975) have recently obtained clear de Broglie interference patterns between neutron beams that have passed through regions of different gravitational potential and are then recombined coherently. Thus, the spatial coordinates of the gravitational potential act empirically in the same way as the spatial coordinates of the electrostatic potential in the usual nonrelativistic treatments of atomic effects. This gives us some confidence in extrapolating macroscopic ideas about space and time, as they occur in the model space of quantum mechanics down to atomic (10^{-8} cm) dimensions.

Granted this, the tests of de Broglie interference patterns performed initially by Davisson and Germer, and repeated for many situations and types of particles (neutral atoms, neutrons, etc., as well as electrons) since, can be invoked as giving evidence close to that of our thought experiment. Typically these are performed not by transmission through slits but by reflection from the surface of a crystal. Since we can use x-rays, calibrated in wavelength against mechanically ruled gratings, to measure the spacing between the atoms on the surface of the crystal, we can argue that the coherent interference from this regular array containing many sites rather than just two is good evidence for the de Broglie wavelength hypothesis. We also can use data from such systems to obtain information analogous to the detectors in the slits in our thought experiment, since we get individual scatterings from individual atomic sites as well as the interference pattern. These are analogous to the single slit patterns of our experiment. We can even perform the analogue of checking this hypothesis by changing the density of detecting particles, since the single atomic patterns (which add incoherently) change with the temperature of the sample in a different way than the blurring that increasing temperature introduces into the coherent interference pattern.

We therefore believe that, although the evidence is indirect, the concept of well-defined geometrical structure and well-defined momenta (subject of course to the restrictions of the uncertainty principle when we try to define both at once) can be self-consistently defined at the atomic level. We think it is probably not often realized, however, that conceptually this implies the existence of particles of much smaller mass and cross section than those under study, if we want to feel comfortable in using all the concepts of the space-time continuum usually assumed without comment in writing down nonrelativistic wave functions. The reason we can get away with this is that we can, by using sufficiently large macroscopic apparatus, define momentum to arbitrarily high precision. This in turn enables us to measure short distances <u>indirectly</u> through quantum mechanical effects which can also be blown up to macroscopic size by going to sufficiently large detecting apparatus. For instance, we might view our double slit experiment as a measurement of the distances w and d. If these were parameters in a molecular model which was free to move under the impact of our beam particles M, rather than being part of the massive baffles which hold the openings under study rigidly in place, the analysis would be less direct but in a reasonably objective sense would still count as a measurement of (quantum mechanical, <u>not</u> classical) distance. From this point of view, the main difference between quantum mechanical and classical model space is that in the quantum mechanical space we can refine either our momentum or our distance measurements as much as we please separately, or sequentially, or accept indirect consequences when our measurement gives some information about each, but not both simultaneously; in the classical model space we are allowed to imagine limiting procedures without these restrictions. But both spaces have limiting procedures and hence conceptually define continua.

The main reason for questioning these limiting procedures, and hence raising doubts about the operational meaningfulness of the space-time concept itself at short distances, comes for me from an effect we have so far not mentioned. This comes from the coupling of relativity to quantum mechanics and was first clearly and succinctly discussed by Wick (1938) in his explanation of the range of nuclear forces given by Yukawa's (1935) meson theory. I have used this insight as one of the starting points in my attempts (Noyes, 1975, 1976, 1977; Note 3) to reformulate elementary particle physics in more operational language. Briefly, the problem is that in a relativistic theory, once one tries to define distances shorter than the Compton wavelength ($\hbar/mc$) of the lightest particle involved, the corresponding momenta are sufficiently energetic to provide the full rest energy (mc^2) of this particle, and additional particles appear in the system. Even if we are well below the energetic threshold for particle creation, the uncertainty principle allows such particles to be present for short times with finite probability. These 'virtual particles' produce indirectly observable effects, such as the nuclear force already alluded to, 'vacuum polarization', etc. Thus, the limiting procedures we have defined above fail at short distance or for sufficiently high momenta.

By now there is considerable experimental evidence that, except for a few absolute conservation laws (charge, baryon number, lepton number, exclusion principle), all known types of particles can be transformed into each other more or less readily at sufficiently high energy. It therefore seems very unlikely that a massive particle less massive than the electron will be discovered in the near future. Thus, in a relativistic quantum theory with this empirical fact added, the units of mass, length, and time are fixed by $\hbar$, c, and m_e, and _the theory is no longer scale invariant_. So far as we can see, this is a powerful argument against using model spaces that allow indefinitely small (continuum) limits to be taken. This is not a new stricture. Bohr and Rosenfeld (1933) analyzed the much more complicated case of the quantum theory of the electromagnetic field in a manner similar in spirit to my presentation here. In that case one has two universal constants (c and $\hbar$) but is allowed to use arbitrarily large masses and charges. Since the theory is scale invariant, they consider themselves at liberty to invoke an arbitrarily complicated apparatus within a wavelength (and it gets pretty complicated) but do finally succeed in rederiving the commutation relations for the electromagnetic field operators. They note in their paper that the argument breaks down in any theory with an intrinsic mass and hence that the same argument cannot be applied to the second quantization of the matter field. In a sense, this current effort of mine can be considered as an addendum to their paper, in which I attempt to give precise operational meaning to nonrelativistic quantum mechanics but end up with the conclusion that this is inevitably frustrated by the Wick-Yukawa mechanism.

References

Bohr, N., & Rosenfeld, L. On the measurability of the electromagnetic field. _Konig Danske Vidensaberns Selskab, Mathematische-Fysik_, 1933, _12_(8).

Bridgman, P. W. _The logic of modern physics_. New York: Macmillan, 1928.

Colella, R., Overhauser, A. W., & Werner, S. A. Observation of gravitationally induced quantum interference. _Physical Review_, 1975, _34_, 1472-1474.

Goldberger, M. L., & Watson, K. _Collision theory_. New York: Wiley, 1964.

Jeffries, H., & Jeffries, B. S. _Mathematical physics_. Cambridge, England: Cambridge University Press, 1950.

Noyes, H. P. Fixed past and uncertain future: A single time covariant quantum particle mechanics. _Foundations of Physics_, 1975, _5_, 37-43 [Erratum: _6_, 125].

Noyes, H. P. A Democritean phenomenology for quantum scattering theory. *Foundations of Physics*, 1976, *6*, 83-100.

Noyes, H. P. A Democritean approach to elementary particle physics. In F. Gilman (Ed.), *Proceedings of the 1976 SLAC Summer Institute*, SLAC 198, November 1976.

Wick, G. C. On the range of nuclear forces in Yukawa's meson theory. *Nature*, 1938, *142*, 993-994.

Yukawa, H. On the interaction of elementary particles, I. *Proceedings of the Physico-Mathematical Society of Japan*, 1935, *17*, 48-57.

Reference Notes

1. Noyes, H. P. *A scientific retrodiction of our past* (Stanford Linear Accelerator Center PUB-1380). Unpublished manuscript, Stanford University, 1974.
2. Noyes, H. P. *Professional responsibility* (Seminar at Stanford for the Social Thought Program). Unpublished manuscript, Stanford University, 1976.
3. Noyes, H. P. *Non-locality in particle physics* (Stanford Linear Accelerator Center PUB-1405, rev.). Unpublished manuscript, Stanford University, 1975.

MEASURING POSITION PROBABILITIES

Nancy Cartwright
Stanford University

I. INTRODUCTION

The Born interpretation provides the fundamental link between theory and observation in quantum mechanics. Schrödinger understood $|\Psi(r)|^2dr$ to be a physical density for charged particles; but this was thought to be inadequate, chiefly because $|\Psi(r)|^2dr$ spreads out in time, whereas the particles themselves remain spatially compact. Born proposed, instead, to take $|\Psi(r)|^2dr$ to represent not a physical, but rather a probability, density. This interpretation is universally accepted in current quantum theory: $|\Psi(r)|^2dr$ gives the probability density that a system in state $\Psi(r)$ is located in a small region around *r*.

Despite its crucial role, it is notoriously difficult to express the Born interpretation coherently, for there is no proper event space for the probabilities to range over. The two-slit experiment and other interference phenomena teach that particles are not located at determinate positions under ordinary circumstances. When a position measurement is performed, however, they are found to be in some place or another. This suggests the usual reading of the Born interpretation: $|\Psi(r)|^2$ represents the probability that the system will be found at *r* if a measurement is made, rather than the probability that the system is at *r*. This reading immediately raises the measurement problem: How can particles which are not located at any determinate position nevertheless be found at some specific place when an observation is made? It also raises a less frequently discussed consistency problem.

If *measurement* in the Born interpretation is to be taken literally, then it must be carried out with a real physical device. But real physical systems affect each

Studies in the Foundations of Quantum Mechanics, pp. 109-118

other only if energy is interchanged between them. A real detector cannot respond to the mere presence of a particle, but will react only if the particle transfers energy to it. If the amplification process is maximally efficient so that the counter registers just in case the appropriate energy is transferred, then a particle will register in the detector if and only if the appropriate energy interchange occurs.

This is the source of the consistency problem. A particle will be found at $\underline{r}$ in a position measurement just in case a detector located at $\underline{r}$ undergoes some specific energy interaction with the particle. The probability of the first event is supposed, under the Born interpretation, to be given by $|\Psi(r)|^2$. But the probability of the second is calculated in a quite different way, using the methods of scattering theory. The Born interpretation is consistent only if the two probabilities can be shown to be equal, or approximately so. Otherwise $|\Psi(r)|^2$ will not give the probability that the system will be found at $\underline{r}$ in a real physical measurement.

On the other hand, it may be the case that <u>measurement</u> in the Born interpretation should not be taken literally. This has been suggested by James L. Park (1968), who distinguishes actual physical measurement (which he calls <u>measurement</u>$_2$), which ought to be treated as an ordinary quantum interaction, from pure observation (<u>measurement</u>$_1$), which might be carried out by a genie riding on the back of the particle. But Park himself concentrates only on measurement$_2$, and, so far as I know, no one has followed up his suggestion and provided an analysis of the phantom measurement$_1$ and how it is to be understood. So long as we continue to interpret measurement as measurement$_2$, carried out with a real physical detector, the consistency problem must be dealt with. This is what I shall do in the following sections.

II. MEASURING POSITIONS

Henry Margenau has long urged that all quantum measurements are ultimately measurements of position (cf. Margenau, 1937, 1958). Position measurements are themselves, however, basically records of scattering interactions. Position measurements occur when the particle whose position is to be measured scatters from a detector. The scattering is inelastic: Energy is not conserved in the particle and the detector is activated by the energy which the particle gives up as it scatters. In the common position-measuring devices--cloud chambers, scintillation counters, and photographic plates--the relevant scattering interaction is the same. The particle scatters from a target in the detector, and the energy which is given up by the particle causes the detector to ionize. The devices differ in how they collect or register the ions; but in each case, the presence of the particle is registered only if the appropriate ionization interaction occurs. So, at best, the probability of a count in the detecting device is equal to the probability of the designated ionization interaction.

Matters may be a lot worse. General background radiation may produce ionization when no particle is present. Conversely, the ion-collecting procedure may be inefficient, and ions which are produced by the scattered particle may fail to register. In principle, it is possible to correct for these effects in calculating the probabilities. For simplicity I shall assume instead that the devices under consideration are ideally efficient, that is, that all and only ions which are produced by the scattered particle are collected and counted. Photographic emulsions are highly efficient in this sense--efficiencies in some cases are greater than 98%. Under this highly idealized assumption, to show consistency requires showing that $|\Psi(\underline{r})|^2$--the Born probability of detecting the particle at $\underline{r}$--is approximately equal to the probability of inelastic scattering into the designated ionization states from a target located at $\underline{r}$.

In fact, this is stronger than necessary, for we are not interested in the absolute values of the probabilities, but merely in their relative values. Consider, for example, the photographic plate, which is the device best suited for determining densities of particles at fairly well-defined positions. In a photographic plate, we are not concerned that the number of spots on the plate should record the actual number of particles, but rather that the pattern on the plate should reflect the distribution of particles. This requires establishing not an equality, but merely a proportionality:

(1)

$$\frac{\text{Probability of ionization in a target at } r}{\text{Probability of ionization in a target at } r'} = \frac{|\Psi(r)|^2}{|\Psi(r')|^2} .$$

For a general justification of the Born interpretation, (1) should be true for any state function for which the problem is well defined. For simplicity I will treat a two-dimensional example, in which a row of detectors is arrayed in a line perpendicular to the $\underline{z}$ axis. We may think of the detectors as the active elements in a photographic plate. In this case, $\Psi(r,t)$ should be arbitrary in shape, except that at $t = 0$ it will be located entirely to the left of the plate (see Fig. 1). It is easiest to establish (1) not in its immediate form, but rather by inverting to get

(2)

$$\frac{\text{Probability of ionization in a target at } r}{|\Psi(r)|^2} = \frac{\text{Probability of ionization in a target at } r'}{|\Psi(r')|^2} .$$

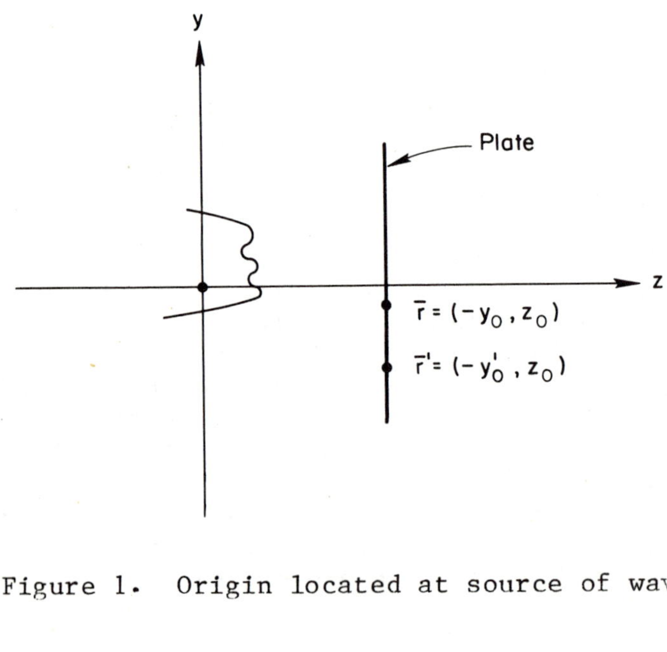

Figure 1. Origin located at source of wave.

Thus, the aim is to show that the ratio in (2) is a constant independent of $\underline{r}$. A shift of coordinates will help. We think of fixing the center of the wave and varying the location of the detector across the plate (Fig. 1). But the same effect is achieved by fixing the location of the detector and varying the center of the wave (Fig. 2). Looked at in this way, the consistency result can be seen as a trivial consequence of a fundamental theorem of scattering theory. This theorem states that the scattering cross section, both total and differential, is a constant independent of the shape or location of the incoming wave packet. The total scattering cross section is essentially the ratio of the total probability for the particle to be scattered divided by the probability for crossing the detector. Roughly, the fundamental theorem assures that the probability of scattering from the detector for a wave centered at $\bar{r}_0$ divided by the probability for "being at" the detector is a constant independent of $\bar{r}_0$. This is just what is required by (2).

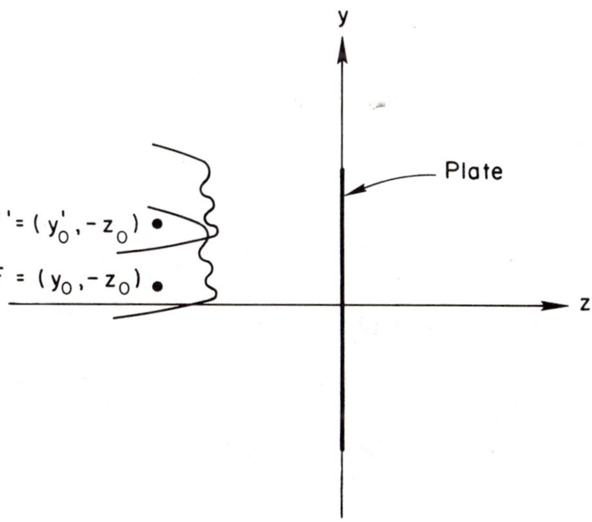

Figure 2. Origin located at detector.

There is one difficulty, however. Standard textbook proofs of this theorem (cf. Gottfried, 1961, par. 12, or Merzbacher, 1961, chap. 11, Sec. 1–3) do not establish the result for arbitrary waves, but only for wave packets which have a narrow spread in the _direction of the momentum_. This is not enough to justify the Born interpretation. For completeness, in the next section, (2) will be calculated directly and shown to hold for arbitrary initial states.

III. THE EQUIVALENCE OF THE TWO DISTRIBUTIONS

Consider a detector located at $y = 0$, $z = 0$, and an arbitrarily shaped wave centered at time $t = 0$ at $\bar{r}_0 = (y_0, -z_0)$:

$$(3) \qquad \Psi(\bar{r},0) = \int f(\bar{k})\exp[i\bar{k}\cdot(\bar{r} - \bar{r}_0)]d\bar{k}$$

and

$$(4) \qquad \Psi(\bar{r},t) = \int f(\bar{k})\exp[ik\cdot(\bar{r} - \bar{r}_0) - i\omega t]d\bar{k}$$

where $\omega = \hbar k^2/2m$. The aim is to show that the ratio of the probability for crossing the detector to the probability for scattering into a designated state (or set of states) is a constant independent of the initial location of the wave along the y axis--i.e., independent of y_0.

Calculating from the Born interpretation, the probability that the system will be counted at some time by a detector at $\bar{r}$ is

$$(5) \qquad \int_{-\infty}^{\infty} |\Psi(\bar{r},t)|^2 dt \ .$$

Introducing $k_x = k \cos\phi$ and $k_y = k \sin\phi$, (5) becomes

$$(6) \qquad \iiint f(k,\phi)f^*(k,\phi')\exp[iky(\sin\phi - \sin\phi') + ikz(\cos\phi - \cos\phi') + iky_0(\sin\phi' - \sin\phi) + ikz_0(\cos\phi - \cos\phi')]k^2 dk\, d\phi'\, d\phi \ ,$$

since $\exp[i(\omega - \omega')t]dt = \delta(\omega - \omega') = \delta(k - k')$. The detector will resolve at best only down to atomic dimensions, so (6) should be integrated over an area around

(0,0) of radius approximately 10^{-8} cm. Performing the dy integral,

(7)

$$\int_{-A}^{A} \exp[iky(\sin\phi - \sin\phi')]dy = \frac{2\sin[kA(\sin\phi - \sin\phi')]}{k(\sin\phi - \sin\phi')}$$

which is of the form $(2A \sin x)/x$. Typically, k will be of the order of 10^{7} cm or less and so $kA \approx 10^{-1}$ cm. Hence, the dy integral will be approximately equal to 2A, since $(\sin x)/x \approx 1$ for small x. Similarly, the dz integral will also be approximately equal to 2A. So the probability for getting a count in the detector as calculated by the Born interpretation is, finally,

(8) $$4A^2 \iiint f^*(k,\phi)f(k,\phi')\exp[ikz_0(\cos\phi - \cos\phi') - iky_0(\sin\phi - \sin\phi')]k^2dk\, d\phi\, d\phi' .$$

To calculate the scattering probability, it is necessary to consider composite states for the system plus detector. I will assume a highly idealized detector in which the sensitive element is a single atom whose outer shell electrons are distinguishable from the measured particles, so that simple theory of inelastic scattering can be used. The treatment in Wu and Ohmura (1962) can be applied directly to our problem, and I shall use their notation to facilitate reference.

Let $\{\xi_i(\bar{r}_2)\}$ represent a complete set of energy eigenstates for the detector when it is not in interaction with the particle, and assume that the detector is initially in the state ξ_n with energy E_n. So long as the particle and the counter do not interact, the state of the composite, for a free particle with momentum $\hbar\bar{k}$ is given by $\Phi_{\bar{k},n}(\bar{r}_1,\bar{r}_2) = \exp[i\bar{k}\cdot(\bar{r}_1 - \bar{r}_0)]\xi_n(\bar{r}_2)$. Under an interaction potential $U(\bar{r}_1,\bar{r}_2)$ which produces outgoing waves in the detected particle, $\Phi_{\bar{k}}$ becomes

(9) $$\Psi'_{\bar{k},n}(\bar{r}_1,\bar{r}_2) = \exp[i\bar{k}\cdot\bar{r}_1]\xi_n(\bar{r}_2) + \sum_m \int (\exp[ik_m r_1]/r_1)a^{\phi}_{k_m}(\theta_1)\xi_m(r_2) ,$$

where sum or integral should be taken depending on whether the state is free or bound. Here $\bar{r}_1 = (r_1, \theta_1)$, and $a^{\phi}_{k_m}(\theta)$ is the scattering amplitude for the detected particle to be scattered from the initial direction $\phi = \arctan k_y/k_x$ into the final direction θ_1 with energy $\hbar^2 k_m^2/2m = \hbar^2 k^2/2m + E_n - E_m$. The states $\Psi'_{\bar{k},n}$ are the asymptotes, for large r_1, of eigenstates for the total Hamiltonian for the interacting systems, corresponding to energy values $(\hbar^2 k^2/2m) + E_n$.

For the case at hand, the measured system is not initially in an eigenstate of momentum, but rather in a superposition of such states with expansion coefficients $\{f(\bar{k})\exp[-ik \cdot r_0]\}$. So $\Phi_n(\bar{r}_1, \bar{r}_2) = \int f(\bar{k}) \Phi_{\bar{k},n}(\bar{r}_1, \bar{r}_2) d\bar{k}$. The corresponding interaction states $\Psi'_n(\bar{r}_1, \bar{r}_2)$ can be expressed as a superposition of the complete set $\Psi'_{\bar{k},n}$ with exactly the same expansion coefficients (cf. Moses, 1955, or Brenig & Haag, 1963). The final state for the particle plus detector is thus, for large r_1,

(10)

$$\Psi'_n(\bar{r}_1, \bar{r}_2) = \int f(\bar{k}) \exp[i\bar{k} \cdot (\bar{r}_1 - \bar{r}_0)] \Phi_n(\bar{r}_2) d\bar{k} + \int f(\bar{k}) \exp(-i\bar{k} \cdot \bar{r}_0) \cdot \left(\sum_m \int [\exp(ik_m r_1)/r_1] a^{\phi}_{k_m}[\theta_1] \xi_m[\bar{r}_2] \right) d\bar{k} .$$

The first term represents that portion of the wave which is not scattered at all and for which the detector is not excited. We will thus be interested only in the second term, $\Psi_{sc}(\bar{r}_1, \bar{r}_2)$.

If the designated state--in the standard devices, an ionization state--to which the counter responds is represented by $\xi_D(\bar{r}_2)$, then for an ideally efficient detector the probability of a count is given by

(11) $$\sum_i \langle \Psi_{sc} | \xi_D \otimes \alpha_i \rangle \langle \xi_D \otimes \alpha_i | \Psi_{sc} \rangle ,$$

where $\{\alpha_i\}$ is a complete orthonormal set of states for the particle. We sum over the final particle states because we are interested only in what state the detector is left in, and not the particle. This is also the justification for using the asymptotic state Ψ_{sc} for which the particle is far from the scattering center. It is possible to sum over a complete set $\{\alpha_i\}$, since those states which are not allowed by energy conservation will make no contribution in the sum.

Ψ_{sc} is of the general form

$$(12)\quad \iint B(k,\phi)\Big(\sum_m \int g_{k_m,\phi}(r_1,\theta_1)\xi_m(r_2)\Big)k\ dk\ d\phi\ ,$$

where $g_{k_m,\phi}(r_1,\theta_1) = \langle \bar{r}_1 | g_{k_m,\phi}\rangle = (\exp[ik_m r_1]/r_1)a_m^{\phi}(\theta)$ and $B(k,\phi) = f(k,\phi)\exp[-iky_0 \sin\phi + ikz_0 \cos\phi]$. Substituting (12) into (11) gives

(13)

$$\iiiint B(k,\phi)B^*(k',\phi')\sum_m \int \sum_{m'} \int g_{k'_{m'},\phi'}\Big|\Big(\sum_i |\alpha_i\rangle\langle\alpha_i|\Big)\Big| g_{k_m,\phi}\rangle$$

$$\langle \xi_{m'} | \xi_D\rangle\langle \xi_D | \xi_m\rangle kk'\ dk\ dk'\ d\phi\ d\phi'\ .$$

Since $\Sigma_i |\alpha_i\rangle\langle\alpha_i| = 1$, and $\langle \xi_{m'} | \xi_D\rangle = \delta_{m'D}$, (13) becomes

(14)

$$\iiiint B(k,\phi)B^*(k',\phi')\langle g_{k'_D,\phi'} | g_{k_D,\phi}\rangle dk\ dk'\ d\phi\ d\phi'$$
$$= \iiint B(k,\phi)B^*(k,\phi') \int a_{k_D}^{\phi}(\theta)a_{k_D}^{\phi'*}(\theta)d\theta\ dk\ dk'\ d\phi\ d\phi'\ ,$$

since $\langle g_{k'_D,\phi'} | g_{k_D,\phi}\rangle = \int r_1^2 a_{k_D}^{\phi}(\theta)a_{k'_D}^{\phi'*}(\theta)(\exp[-ik_D r_1]/r_1) \cdot (\exp[ik_D r_1]/r_1)r_1^2 dr\ d\theta = \delta_{k,k'}$. In the usual case, however, the scattering amplitude $a_k^{\phi}(\theta)$ is slowly varying in $\bar{k}$. To the extent that this is true, $a_{k_D}^{\phi}(\theta)$ and $a_{k_D}^{\phi'}(\theta)$ can be replaced approximately by the mean amplitude $a_0(\theta)$; and the probability for the detector to be excited to the state ξ_D will be

$$(15)\quad \iiint f^*(k,\phi)f(k,\phi')\exp[ikz_0(\cos\phi - \cos\phi') - iky_0(\sin\phi - \sin\phi')]k^2\ dk\ d\phi\ d\phi'\ .$$

From (15) and (18) we see that the ratio of the Born probability for "finding the particle" compared to the probability of getting a count in the detector is just

$$(16)\quad \frac{\text{Probability for detection calculated via Born interpretation}}{\text{Probability for detection calculated via scattering theory}} = \frac{4A2}{\int |a_0(\theta)|^2 d\theta}\ .$$

This is indeed a constant independent of y_0, as required.

References

Brenig, W., & Haag, R. General quantum theory of collision processes. In M. Ross (Ed.), Quantum scattering theory. Bloomington: Indiana University Press, 1963.

Gottfried, K. Quantum mechanics (Vol. 1). New York: Benjamin, 1961.

Margenau, H. Critical points in modern physical theory. Philosophy of Science, 1937, 4, 352-356.

Margenau, H. Philosophical problems concerning the meaning of measurement in physics. Philosophy of Science, 1958, 25, 23ff.

Merzbacher, E. Quantum mechanics (2nd ed.). New York: Wiley, 1961.

Moses, H. E. The scattering operator and the adiabatic theorem. Nuovo Cimento, 1955, 1, 104.

Park, J. L. Quantum theoretical concepts of measurement: Part II. Philosophy of Science, 1968, 35, 389-411.

Wu, T., & Ohmura, T. Quantum theory of scattering. Englewood Cliffs, N.J.: Prentice-Hall, 1962.

LOCALITY AND THE ALGEBRAIC STRUCTURE OF QUANTUM MECHANICS

William Demopoulos
University of Western Ontario

Bell (1966) first raised the issue of locality in a highly influential analysis of the relevance of several theorems for the problem of hidden variables--that is, the problem of whether there exists a physically motivated deterministic interpretation of the statistical relations of the quantum theory. The principal result considered by Bell was the corollary to Gleason's theorem that there are no two-valued measures on the partial Boolean algebra (or partially ordered set) L(H) of closed linear subspaces of a separable Hilbert space of three or more dimensions. It follows from this corollary that there are no dispersion-free states on the set of physical magnitudes represented by the self-adjoint linear operators on H. Bell observed that this result excludes deterministic interpretations that attempt to preserve a certain amount of the algebraic structure of the magnitudes. Specifically, there must exist a homomorphism h: L(H) -> Z_2 from L(H) onto the two-element Boolean algebra Z_2 which preserves the algebraic operations on <u>compatible</u> subspaces of L(H), or, equivalently, on the corresponding commuting projection operators acting on H. To preserve this structure implies certain relations between incompatible magnitudes. For example, if a projection operator P occurs in the spectral representation of two nondegenerate linear operators A and B, then the proof of the corollary to Gleason's theorem requires that the image of P under h is the same whether P is regarded as a function of A or as a function of B. Bell argued that since A- and B- measurements might interfere with one another, there is no a priori reason to impose this requirement. According to Bell, such a condition would be justified in certain special cases. For example, given a pair of spin-1/2 particles in the

Studies in the Foundations of Quantum Mechanics, pp. 119-144

singlet spin state, the value of a spin magnitude referring just to S should not depend on the kind of magnitude considered at S′.

Bell called the requirement that functional relations of this type be preserved in the case of coupled systems, such as S + S′, a condition of locality; a hidden variable theory satisfying this condition is then a local hidden-variable theory. Given a principle regarding the computation of certain joint probabilities, Bell (1964) derived an inequality that every local hidden variable interpretation of the singlet spin state statistics must satisfy, but this inequality is violated quantum mechanically.

It is generally believed that locality is a weaker requirement than the homomorphism condition--the condition that there exist a homomorphism from L(H) onto Z_2--and that therefore the probabilistic analysis of S + S′ given by Bell cannot be replaced by one that is purely algebraic. That is, it is generally believed: (i) while locality is physically motivated, the homomorphism condition is not, and (ii) no algebraic analysis of a pair of spin-1/2 particles that restricts itself to the condition of locality can exclude a hidden variable interpretation of the quantum statistics.

The aim of this paper is to show that locality by itself suffices to exclude a hidden variable interpretation because <u>locality implies the homomorphism condition</u>. To prove this we adapt Maczyński′s (1971) notion of a Boolean representation of a family of magnitudes to the spin magnitudes of S + S′. A Boolean representation of this family of magnitudes exists if and only if the condition of locality is satisfied, and the spin magnitudes of S + S′ have a Boolean representation only if there is a homomorphism from a certain $L \subset L(H^4)$ into a Boolean algebra. Applying a theorem of Kochen and Specker′s (1967), we infer that there is no local hidden-variable theory of a finite subset of this set of spin magnitudes.

The discussion assumes familiarity with the concepts of a partial algebra and partial Boolean algebra introduced by Kochen and Specker. With minor exceptions, the terminology and notation of this paper is consistent with Maczyński (1971); that paper should be consulted for the exposition and justification of the mathematical framework employed here.

I. NOTATION AND TERMINOLOGY

We shall be almost exclusively concerned with a composite system $S + S'$ consisting of a pair of spin-1/2 particles that we may suppose to have interacted and then separated. Quantum mechanically, the spin magnitudes of $S + S'$ are represented by the partial algebra $\underline{A} \otimes \underline{A}'$ of self-adjoint linear operators over the tensor product $H \otimes H'$ of the two-dimensional Hilbert spaces associated with S and S′. $L = L(H \otimes H')$ denotes the partial Boolean algebra of subspaces of $H \otimes H'$.

Within the mathematical framework presented by Maczyński, the relevant magnitudes of $S + S'$ may be described as follows: Each maximal (nondegenerate) spin magnitude $A \otimes B \in \underline{A} \otimes \underline{A}'$ is represented by a σ-homomorphism $\mu_{AB}: B(R^2) \to L$ from the Borel subsets of the Cartesian product of the reals into L. (To simplify the notation, ⊗ has been replaced by juxtaposition whenever $A \otimes B$ appears as an index.)

Now let

$$\{\lambda_1, \lambda_2, \lambda_3, \lambda_4\}$$

be the set of points that comprise the spectrum of $A \otimes B$; it consists of the pairs of real numbers:

$$(1/2, 1/2),\ (1/2, -1/2),\ (-1/2, 1/2),\ (-1/2, -1/2)$$

--in that order, and in units $\hbar$. We denote the one-dimensional subspace of L that is the image of $\{\lambda_i\}$ under μ_{AB} by ab_i. The image of $E \in B(R^2)$ under μ_{AB} is defined by the condition

$$\begin{aligned}\mu_{AB}(E) = \mu_{AB}(\{\lambda_{k_1}\}) + \mu_{AB}(\{\lambda_{k_2}\}) \\ + \mu_{AB}(\{\lambda_{k_3}\}) + \mu_{AB}(\{\lambda_{k_4}\}) ,\end{aligned}$$

where

$$\{\lambda_{k_1}, \lambda_{k_2}, \lambda_{k_3}, \lambda_{k_4}\} = E \cap \{\lambda_1, \lambda_2, \lambda_3, \lambda_4\} .$$

If this set is empty, $\mu_{AB}(E) = \Lambda_L$.

Spin magnitudes of the form $A \otimes I$, where I is the identity operator on H', are called <u>local</u> spin magnitudes. $A \otimes I$ is represented by the σ-homomorphism $\mu_{AI}: B(R^2) \to L$ defined by

$$\mu_{AI}(E) = \mu_{AI}(\{\eta_{k_1}\}) + \mu_A(\{\eta_{k_2}\}) ,$$

where

$$\{\eta_{k_1}, \eta_{k_2}\} = E \cap \{\eta_1, \eta_2\}$$

and

$$\eta_1 = (1/2, 1) ,$$

$$\eta_2 = (-1/2, 1)$$

are the points comprising the spectrum of $A \otimes I$. Again, if $E \cap \{\eta_1, \eta_2\}$ is empty, $\mu_{AI}(E) = \Lambda_L$. The image of $\mu_{AI}(\{\eta_j\})$ is a two-dimensional subspace of $H \otimes H'$, denoted by ai_j. The σ-homomorphism μ_{IA} corresponding to the local spin magnitude $I \otimes A$ is defined in the obvious way.

We let L_{AB} denote the range of μ_{AB}, L_{AI} the range of μ_{AI}, etc. L_{AB} is a 16-element Boolean subalgebra of L; L_{AI}, a four-element Boolean subalgebra of L.

We remark that

$$\mu_{AI}(\{\eta_1\}) = ai_1 = ab_1 + ab_2 = \mu_{AB}(\{\lambda_1\}) + \mu_{AB}(\{\lambda_2\})$$

and

$$\mu_{AI}(\{\eta_2\}) = ai_2 = ab_3 + ab_4 = \mu_{AB}(\{\lambda_3\}) + \mu_{AB}(\{\lambda_4\})$$

for any B referring to S′. For, given any state α of S + S′, the probability that the spin of S in the direction A is positive is equal to the probability that the spin of S in the direction A is positive and the spin of S′ in the direction B is positive or negative.

Similarly, putting $\eta'_1 = (1, 1/2)$, $\eta'_2 = (1, -1/2)$,

$$\mu_{IB}(\{\eta_1\}) = ib_1 = ab_1 + ab_3 = \mu_{AB}(\{\lambda_1\}) + \mu_{AB}(\{\lambda_3\})$$

and

$$\mu_{IB}(\{\eta_2\}) = ib_2 = ab_2 + ab_4 = \mu_{AB}(\{\lambda_2\}) + \mu_{AB}(\{\lambda_4\}) ,$$

for any A referring to S. Thus $L_{AI}, L_{IB} \subset L_{AB}$, and therefore there are Borel functions f,g such that $A \otimes I = f(A \otimes B)$ and $I \otimes B = g(A \otimes B)$. These facts will be used in the proof of Theorem 2.1.

II. BOOLEAN REPRESENTATIONS OF PHYSICAL MAGNITUDES AND LOCALITY

Let K be the set of all local spin magnitudes, and M the set of all maximal spin magnitudes. Put $P = K \cup M$. Since the elements of P are represented by homomorphisms from $B(R^2)$ into L, P is said to be <u>based on</u> L. Roughly speaking, this section investigates the possibility of replacing L with a Boolean algebra.

A pair $(B, \{r_{AI}\}_{AI \in K} \cup \{r_{AB}\}_{AB \in M})$ is said to be a <u>Boolean representation</u> for P if:

(1) B is a σ-complete Boolean algebra and for every $A \otimes I \in K$ and $A \otimes B \in M$, r_{AI} and r_{AB} are σ-homomorphisms from $B(R^2)$ into B such that the union of the ranges of r_{AI} and r_{AB} σ-generates B.

(2) Let $ai \in L_{AI}$, $ab \in L_{AB}$, with $\mu_{AI}(E) = ai$ and $\mu_{AB}(F) = ab$. Then for each $A \otimes I \in K$ and $A \otimes B \in M$, the corresponding maps $h_{AI}: ai \mapsto r_{AI}(E)$ and $h_{AB}: ab \mapsto r_{AB}(F)$ are σ-homomorphisms of L_{AI} and L_{AB} into B.

(3) If for $C \otimes I$, $A \otimes B \in P$ and a Borel function f we have $C \otimes I = f(A \otimes B)$, then $r_{CI} = f(r_{AB})$, that is, $r_{AI} = r_{AB} \circ f^{-1}$.

Following Maczyński we shall say that P is Boolean representable if there exists a Boolean representation of P. Clearly, P is Boolean representable if there exist maps $A \otimes B \mapsto r_{AB}$, $A \otimes I \mapsto r_{AI}$ from P into the set of magnitudes based on a Boolean σ-algebra B such that all functional relations between the elements of P are preserved--the range of $A \otimes B$ being mapped onto the range of r_{AB}, and the range of $A \otimes I$ onto the range of r_{AI}. Hence a Boolean representation of P preserves the inclusion relations that hold between the ranges of the magnitudes in P: That is, if $\phi: L_{AI} \to L_{AB}$ is the inclusion homomorphism from L_{AI} into L_{AB}, we have $h_{AB} \circ \phi = h_{AI}$.

At this stage I wish to consider certain general objections to the usefulness of the notion of a Boolean representation of a set of magnitudes for the problem of hidden variables. This discussion will motivate the introduction of locality.

Let Q be an arbitrary set of magnitudes of some quantum mechanical system, S the set of its statistical states, and B(R) the Borel subsets of the real line; $p\colon Q \times S \times B(R) \to [0,1]$ is the probability map described in Maczyński (1971). Then the sense in which the spectral representation of the elements of Q requires a functional relation to hold is probabilistic: If $f\colon R \to R$ is a Borel function, $A = f(B)$ means

$$p(A,\alpha,E) = p(B,\alpha,f^{-1}(E))$$

for all $E \in B(R)$ and all $\alpha \in S$. Of course, we might also have

$$A = g(C)$$

with C incompatible with B.

Let us call a representation of Q <u>deterministic</u> if it admits dispersion-free states on Q. In order to justify the requirement that a deterministic representation of Q must be a <u>Boolean</u> representation--that is, that the representation must preserve functional relations--Kochen and Specker interpret $A = f(B)$ and $A = f(C)$ as relations between the values of A and B and A and C. That this extends what is guaranteed by the spectral theorem has been noted by Bell (1966, p. 451) and also by Gudder (1970, p. 435). They argue that since a measurement of B interferes with a measurement of C, there is no reason to accept this interpretation of functional relations.

Our present concern is solely with functional relations between the local and maximal spin magnitudes of S + S′. In the case of such correlated systems it seems that more than the assertion of the probabilistic equivalence of certain statements about S + S′ is justified. We have

$$ai_1 = ab_1 + ab_2 = ac_1 + ac_2 \ ,$$

with AI ♀ AB, AI ♀ AC, and AB and AC incompatible. But now we may argue as follows: Since S and S′ are spatially separated, the value of the magnitude $A \otimes I$ cannot be affected by the kind of magnitude considered at S′, so that in this case the incompatibility of $A \otimes B$ and $A \otimes C$ cannot be understood in terms of the interference of $A \otimes B$- and $A \otimes C$- measurements. But this removes the only objection to treating $f(A \otimes B) = A \otimes I = g(A \otimes C)$ as a functional relation between the values of $A \otimes B$, $A \otimes I$, $A \otimes C$. Thus, we may reasonably require of any deterministic representation of the magnitudes of S + S′ that $r_{AI} = r_{AB} \circ f^{-1}$. In other words, Bell′s condition of locality leads immediately to Boolean representability--at least in the case of spatially separated composite systems.

To sum up this brief digression: In the case of a single system, Bell and Gudder will allow only the following constraint. For $\alpha \in S$, let m_α be the corresponding measure on the direct product representation of all the magnitudes $A \in Q$. Recall, this is the representation that treats the magnitudes in Q as independent random variables defined over a common probability space. (See Kochen & Specker, 1967, Sec. 1, and Wigner, 1970. This will be discussed in greater detail below.) Now let a be an element of L_A, L_B, L_C, with B and C incompatible. Then according to Bell and Gudder we may require of the representation of Q that

$$m_\alpha(h_A(a)) = m_\alpha(h_B(a)) = m_\alpha(h_C(a)) \; ;$$

we must not require that

$$h_A(a) = h_B(a) = h_C(a) \; .$$

But in the case of spatially separated, composite systems, Bell at least is willing to allow the stronger condition, since this follows from general considerations regarding the existence of approximately isolated systems and the local character of physical laws. Thus, in this special case, locality coincides with the condition that the representation of P must be a Boolean representation.

It is important to notice that even though a set of magnitudes is based on L, their Boolean representability need not require the existence of a homomorphism from L into a Boolean algebra. For example, the set M of maximal magnitudes may be based on L; M is also Boolean representable (Maczyński, 1971, Theorem 4.5). M even generates the algebra of all magnitudes based on L. But the set of all quantum mechanical magnitudes is not Boolean representable, for this does imply the existence of a homomorphism from L into a Boolean algebra (Maczyński, 1971, Theorem 4.1). An examination of Maczyński's proof of this fact shows that the set M of all maximal magnitudes together with the set N of all magnitudes A such that L_A is a four-element Boolean subalgebra of L--the set of 'minimal magnitudes'--is not Boolean representable. The local spin magnitudes of S + S′ form a proper subset of the set of minimal magnitudes. We may therefore ask: Is the set of local and maximal spin magnitudes Boolean representable? This question is partially answered by the following theorem.

THEOREM 2.1. <u>If the set</u> $K \cup M$ <u>of local and maximal spin magnitudes of</u> S + S′ <u>is Boolean representable, there is a homomorphism</u> h: L -> B <u>from</u> L <u>into a Boolean algebra</u> B.

<u>Proof</u>. Let $(B, \{r_{AI}\}_{AI \in K} \cup \{r_{AB}\}_{AB \in M})$ be a Boolean representation of $K \cup M$. We shall define a homomorphism from L into B. Let L_{AB} be an arbitrary maximal Boolean subalgebra of L. We note that L_{AB} has 2^{2^2} elements, and so is a <u>free</u> Boolean algebra, with a set of two free generators. (See Sikorski, 1964, Sec. 14.) The sets of free generators of L_{AB} consist of all pairs of non-orthogonal two-dimensional subspaces in L_{AB}. Now the four-element Boolean algebras L_{AI} and L_{IB} are included in L_{AB} and each is generated by a distinct two-dimensional subspace. Moreover, the subspaces are not orthogonal. Thus, any set consisting of an atom of L_{AI} and an atom of L_{IB} is a set of free generators of L_{AB}. It is clear that the same analysis applies to any maximal Boolean subalgebra of L. Let L′ be the union of the ranges of the local spin magnitudes. Define a map h: L′ -> B by putting h(a) = $h_i(a)$ for some $h_i: L_i$ -> B where L_i is the range of a local or maximal spin magnitude that contains a. Notice, if L_j and L_k are two maximal Boolean subalgebras such that $L_j \cap L_k$ contains $a \in L'$, then a belongs to some <u>local</u> subalgebra $L_i \subset L_j \cap L_k$. Whence, by condition (3), $h_j(a) = h_i(a) = h_k(a)$, and so h is well defined on L′. Thus, h carries a set of generators of each maximal Boolean subalgebra into B, and the image of $a \in L'$ under h does not depend on the choice of Boolean subalgebra to

which it belongs. We shall now extend h to L. Let $\phi(a_1,\ldots,a_n)$ be a Boolean polynomial in the elements $a_1,\ldots,a_n \in L'$. Put

$$h(\phi(a_1,\ldots,a_n)) = \phi(h(a_1),\ldots,h(a_n)) .$$

Since every element of L is expressible as a Boolean polynomial of some pair of elements a_i,a_j of L', h is defined on L. To show that h is a map we proceed as follows. Let a be any element of L, and let $\phi(a_1,a_2) = \psi(b_1,b_2) = a$ be two representations of a in terms of the free generators $a_1,a_2 \in L_j$, $b_1,b_2 \in L_k$, where L_j and L_k are maximal Boolean subalgebras of L. It is clear from the analysis of Section I that a_1,a_2,b_1,b_2 may always be chosen so that each is the image of a local spin magnitude. By the reflexiveness of $\overset{\circ}{+}$,

$$\phi(a_1,a_2) \overset{\circ}{+} \psi(b_1,b_2) ,$$

hence

$$\phi(a_1,a_2) \wedge \psi(b_1,b_2)^{\perp} = \Lambda_L .$$

Since $\Lambda_L \in L'$,

$$h(\phi(a_1,a_2) \wedge \psi(b_1,b_2)^{\perp}) = h(\Lambda_L) = \Lambda_B .$$

So by the definition of h,

$$\phi(h(a_1),h(a_2)) \wedge \psi(h(b_1),h(b_2))^{\perp} = \Lambda_B .$$

By the same argument,

$$\begin{aligned} h(\phi(a_1,a_2)^{\perp} \wedge \psi(b_1,b_2)) \\ &= \phi(h(a_1),h(a_2))^{\perp} \wedge \psi(h(b_1),h(b_2)) \\ &= \Lambda_B . \end{aligned}$$

Thus, h is well defined on L. It is easily verified that h is a homomorphism from L into B.

The preceding argument shows that h is a homomorphism on the partial Boolean algebra $L = \bigcup \{L_i : i \in I\}$ where each L_i is the range of a maximal spin magnitude or an arbitrary Boolean subalgebra of the range of a maximal spin magnitude. The restriction $♀ \cap L \times L$ of the compatibility relation of $L(H \otimes H')$ to L properly contains the compatibility relation of L. For example, if the components of spin $A \otimes A$, $B \otimes B$ refer to orthogonal directions α and β in E^3, the planes

$$a = \mu_{AA}(\{(1/2,-1/2)\}) \vee \mu_{AA}(\{(-1/2,1/2)\})$$

$$b = \mu_{BB}(\{(1/2,-1/2)\}) \vee \mu_{BB}(\{(-1/2,1/2)\})$$

are compatible in $L(H \otimes H')$, but do not belong to a common L_i of $\{L_i: i \in I\}$. Hence L is not a partial Boolean subalgebra of $L(H \otimes H')$. Of course, wherever the partial operations are defined on L, they coincide with the operations of $L(H \otimes H')$. Hence, we may consider the partial Boolean subalgebra L^* of $L(H \otimes H')$ generated by L. Our problem is to extend $h: L \to B$ to a homomorphism $h^*: L^* \to B$.

For $a \in L$, let $h^*(a) = h(a)$, and in general, for any Boolean polynomial $\phi(a_1,\ldots,a_n)$ in the elements $a_1,\ldots,a_n \in L$, let

$$h^*(\phi(a_1,\ldots,a_n)) = \phi(h(a_1),\ldots,h(a_n)) .$$

Since $\Lambda_{L^*} = \Lambda_L$, h^* is defined for Λ_{L^*} and we have

$$h^*(\Lambda_{L^*}) = h(\Lambda_L) = \Lambda_B .$$

Let us consider the action of h^* on the ray H_1--the one-dimensional subspace of $L(H \otimes H')$ containing the singlet spin state

$$\begin{aligned}\psi &= 2^{-1/2}[\alpha_+ \otimes \alpha_- - \alpha_- \otimes \alpha_+] \\ &= 2^{-1/2}[\beta_+ \otimes \beta_- - \beta_- \otimes \beta_+] ,\end{aligned}$$

where $\alpha_+ \otimes \alpha_-$ is the eigenvector corresponding to the eigenvalue $(1/2,-1/2)$ of $A \otimes A$, etc., and $A \otimes A$, $B \otimes B$ refer to orthogonal directions of E^3. Then

$$\begin{aligned}H_1 &= [\mu_{AA}(\{1/2,-1/2)\}) \vee \mu_{AA}(\{(-1/2,1/2)\})] \\ &\quad \wedge [\mu_{BB}(\{(1/2,-1/2)\}) \vee \mu_{BB}(\{(-1/2,1/2)\})] \\ &= a \wedge b .\end{aligned}$$

Let a', b' be two possibly different planes such that $H_1 = a' \wedge b'$. Then we must show

$$h^*(a \wedge b) = h^*(a' \wedge b') .$$

Now

$$(a \wedge b) = H_1 = (a' \wedge b') ,$$

so that

$$(a \wedge b) \wedge (a' \wedge b')^{\perp} = \Lambda_{L^*} .$$

Since

$$\Lambda_{L^*} = \Lambda_L \in L ,$$

$$h^*((a \wedge b) \wedge (a' \wedge b')^{\perp}) = h^*(\Lambda_{L^*}) = h(\Lambda_L) = \Lambda_B .$$

Hence by the definition of h^*,

$$h^*((a \wedge b) \wedge (a' \wedge b')^{\perp})$$

$$= ((h(a) \wedge h(b)) \wedge (h(a') \wedge h(b'))^{\perp}$$

$$= \Lambda_B .$$

But

$$h(a) \wedge h(b) = h^*(a \wedge b)$$

and

$$h(a') \wedge h(b') = h^*(a' \wedge b') .$$

Thus,

$$h^*(a \wedge b) \wedge h^*(a' \wedge b')^{\perp} = \Lambda_B$$

and therefore

$$h^*(a \wedge b) \leq h^*(a' \wedge b') .$$

Similarly,

$$h^*(a' \wedge b') \leq h^*(a \wedge b) ,$$

and we have shown

$$h^*(a \wedge b) = h^*(a' \wedge b') .$$

Thus, h^* is well defined on H_1.

Let $H_3 = H_1^{\perp}$. Put $h^*(H_3) = h^*(H_1)^{\perp}$. Since h^* is defined on the two-dimensional subspaces of the form

$$a = \mu_{AA}(\{(1/2,-1/2)\} \vee \mu_{AA}(\{(-1/2.1/2)\})$$
$$= (aa_2 \vee aa_3)$$

it is defined on their orthogonal complements

$$a^{\perp} = \mu_{AA}(\{(1/2,1/2)\}) \vee \mu_{AA}(\{(-1/2,-1/2)\})$$
$$= (aa_1 \vee aa_4) \ .$$

Therefore h^* is defined on the lines

$$H_3 - a^{\perp} = H_3 \wedge a$$

of H_3. The verification that h^* is well defined on all of L^* and a homomorphism is straightforward. Thus, we have established the following proposition:

> If $K \cup M$ is Boolean representable, there is a homomorphism $h^*: L^* \to B$.

I shall now briefly indicate the significance of our conclusions regarding h^* and H_3. A full discussion will be presented in another paper.

Notice first that the subspaces aa_1, $H_3 \wedge a$, aa_4 are spanned by the vectors $\alpha_+ \otimes \alpha_+$, $2^{-1/2}[\alpha_+ \otimes \alpha_- + \alpha_- \otimes \alpha_+]$, $\alpha_- \otimes \alpha_-$ (respectively). Taken in this order, these vectors correspond to the eigenvalues $\hbar, 0, -\hbar$ of the projection PJ_α of the operator J_α onto H_3. PJ_α represents the component of spin in the direction α of E^3 for a spin-1 system in the triplet spin spate. Similarly, $\beta_+ \otimes \beta_+$, $2^{-1/2}[\beta_+ \otimes \beta_- + \beta_- \otimes \beta_+]$, $\beta_- \otimes \beta_-$ correspond to the operator PJ_β, etc. <u>Indeed, it is possible to show that L^* contains the entire partial Boolean algebra of the three-dimensional representation of a spin-1 system.</u> (I am indebted to Simon Kochen for this observation.) It follows that there is no homomorphism $\phi: L^* \to Z_2$ such that $\phi(H_3) = 1$, for this would imply the existence of a Z_2-valued homomorphism on $B(E^3)$, and hence on the finite partial Boolean subalgebra D of $B(E^3)$, contradicting Kochen and Specker (cf. Theorem 1, and Sec. 4 [The Operators as Observables]). Therefore L^* is not <u>weakly</u>

<u>embeddable</u> into a Boolean algebra (i.e., there is no homomorphism h: L* -> B such that h is one-one on compatible elements of L*) and this result depends on only a finite subalgebra of L*. Combining this with theorems of Dwinger (1967, p. 323) and Maczyński (1971, Theorem 4.2) on direct limits of Boolean algebras and Boolean representations, we have the following corollary to Theorem 2.1:

COROLLARY 2.2. <u>There is no Boolean representation of</u> $K \cup M$.

It is instructive to compare this characterization of local deterministic representations of P with Wigner's elegant analysis of the derivation of Bell's inequality.

A representation of P that satisfies just conditions (1) and (2) of the definition of a Boolean representation will be called a <u>product</u> representation of P. Wigner begins his discussion by observing that every family of magnitudes has such a product representation; moreover, any set of probability measures is recoverable within a product representation by simply defining a measure m_ψ on B to be the measure induced by the product of the distribution functions associated with each of the magnitudes in P by the statistical state ψ. Wigner considers spin magnitudes of S + S′ associated with just three directions in space. There are nine maximal magnitudes: $A \otimes A$, $A \otimes B$, ..., $C \otimes C$ (sometimes denoted $Q_1, \ldots, Q_9$, respectively), each with four possible values λ_1, λ_2, λ_3, λ_4. In the product representation there are 4^9 relevant regions of the space $X = \Pi_{i=1}^{9} X_i$ corresponding to the 4^9 sequences of possible values. (Each X_i is the phase space of the i-th magnitude Q_i.)

Wigner is concerned with the six local spin magnitudes $A \otimes I$, $B \otimes I$, ..., $I \otimes C$, each of which has two possible values η_1, η_2 (or η_1', η_2'). In the product representation, each local spin magnitude has three representatives. (For example, $A \otimes I$ appears as a function of $A \otimes A$, $A \otimes B$, and $A \otimes C$.) Without locality, the value of $A \otimes I$ depends on whether $A \otimes I$ is regarded as $f(A \otimes A)$, $g(A \otimes B)$, $h(A \otimes C)$--in other words, the value of $A \otimes I$ depends on the choice of representing function $r_{AA} \circ f^{-1}$, $r_{AB} \circ g^{-1}$, $r_{AC} \circ h^{-1}$ of the local spin magnitude $A \otimes I$. Wigner observes that by locality there are only 2^6--not 4^9--essentially different regions of X corresponding to the union of the images of the maps $r_{Q_1} \circ f_1^{-1}, \ldots, r_{Q_9} \circ f_9^{-1}$. But this is clearly equivalent to identifying different representatives of a given local spin magnitude and therefore agrees with our reconstruction of locality as Boolean representability.

There are two major differences between our analysis and Wigner's. To state them we require some additional notation. Let α be the subset of X associated with the value of a local spin magnitude referring to S, and β the subset associated with a local spin magnitude referring to S′. Let m_ψ be the product measure on B corresponding to the statistical state ψ, so that the projection of m_ψ into the i-th factor of B reproduces the probabilities of the possible outcomes for the magnitude Q_i. To derive Bell's inequality, Wigner must assume that the joint probability of the events represented by α and β is simply $m_\psi(\alpha \cap \beta)$. Wigner shows that in this finite example --one consisting of only nine maximal spin magnitudes and six local spin magnitudes--if ψ is the singlet spin state, then there is no m_ψ that reproduces the statistical relations determined by ψ. But to show this, Wigner requires that the product representation must compute the probability of certain joint events according to the principle just noted. Our derivation of the impossibility of a local deterministic representation of P makes no such assumption. Thus, in this sense, Theorem 2.1 supports a much stronger conclusion than the theorem that m_ψ must satisfy the Bell inequality.

The second departure from Wigner derives from the fact that we have employed <u>all</u> of the spin magnitudes in $K \cup M$ to exclude local hidden variables, while Wigner makes do with a finite number. But this is not so important a difference as it might at first appear. For we may apply Kochen and Specker's Theorem 1 and infer that there is no local representation of the spin system S + S′--and this result depends on only a <u>finite</u> subalgebra of the partial algebra of spin magnitudes. Thus, our conclusion is as finitary as Wigner's.

We have just noted that while our argument is completely algebraic and depends solely on locality, Wigner requires an assumption concerning the computation of certain joint probabilities. This bears on a number of discussions of the Bell-Wigner analysis of hidden variables. To take just one example, Fine (1975) introduces what we have called a product representation of the spin magnitudes of $S + S'$. Fine claims that this representation may be modified so that it satisfies locality--what cannot be satisfied is the condition Bell and Wigner impose on joint probabilities. (Cf. Fine, 1975, p. 269.) Fine therefore drops this principle from his account and concludes that there are local representations of the spin system $S + S'$ that admit two-valued measures. This is certainly correct if the account is restricted to the spin magnitudes associated with just three directions in physical space. But the argument of this paper shows that Fine's analysis cannot be extended to include all of the spin magnitudes belonging to $S + S'$. Indeed, by the finitary theorem of Kochen and Specker, Fine's analysis cannot be extended to certain finite subsets of P. (Any set of spin magnitudes whose ranges generate a finite partial Boolean algebra that contains the rays used in the proof of Corollary 2.2 will do.) Exactly the same remarks apply to the analysis of Suppes and Zanotti (1975).

III. LOCAL PROBABILITY MEASURES

In this section we shall show that the argument of Section II goes through under a certain weakening of the condition of locality. Our discussion depends on a theorem of Maczyński's relating Boolean representations of families of magnitudes to the direct limit of the Boolean subalgebras of L which correspond to their ranges. The relevant notions are briefly summarized here.

Let $N = \{L_i: i \in I\}$ be the family of Boolean subalgebras of L belonging to the range of a local or maximal spin magnitude. We note that N has a natural partial order determined by the substructure relation, and we assume that the indexing by I is consistent with this order. Let ϕ_{ij}, $i,j \in I$, denote the inclusion homomorphism from L_i into L_j. Then the direct limit of N is the pair $(B,\{h_i: i \in I\})$ where B is a (possibly degenerate--i.e., $\wedge = \vee$) Boolean σ-algebra and for each $i \in I$, $h_i: L_i \rightarrow B$ is a σ-homomorphism such that

(i) $h_i = h_j \circ \phi_{ij}$ for $L_i \subset L_j$

and

(ii) for every pair $(C,\{h_i: i \in I\})$, where C is a Boolean σ-algebra and $h_i: L_i \to C$ is a σ-homomorphism, there is a unique homomorphism $h: B \to C$ such that $h_i = h \circ h_i$ for all $i \in I$.

Maczyński has shown that the family of ranges of a set of magnitudes has a nondegenerate direct limit if and only if the magnitudes are Boolean representable (1971, Theorem 4.2). (The relation between direct limits and Kochen and Specker's notion of a homomorphic relation was investigated in Demopoulos, 1976.)

Let h_i, h_j be homomorphisms carrying (respectively) a local and a maximal Boolean subalgebra $L_i, L_j \in N$, into B, where B is a product representation of the local and maximal spin magnitudes of $S + S'$. Call a probability measure m on B a <u>local probability measure</u> if $m(h_i(a) \dot{-} h_j(a)) = 0$ for all $a \in L_i$, and all L_i and L_j such that $L_i \subset L_j$. (Thus, we require only that the <u>measure</u> of the symmetric difference of different representatives of a should be zero.)

THEOREM 3.1. <u>Let</u> B <u>be a product representation of</u> P. <u>Then there are no nontrivial local probability measures on</u> B.

<u>Proof</u>. We observe that if m is a nontrivial local probability measure on B, the ideal J generated by all elements of the form

$$h_i(a) \dot{-} h_j(a)$$

is proper. The quotient B/J of B by J, together with the family $\{k_i: i \in I\}$ of maps k_i such that $k_i = \phi | L_i$, where ϕ is the canonical embedding of B onto B/J, is the direct limit of $\{L_i: i \in I\}$ (cf. Dwinger, 1967). Since J is proper, $(B/J, \{k_i: i \in I\})$ is nondegenerate. Hence, by Maczyński's Theorem 4.2, P is Boolean representable. But this contradicts the fact that P is not Boolean representable and proves the theorem.

<u>Remark</u>. In the course of the proof of the above theorem we have established the following proposition:

> If a product representation of P admits a nontrivial local probability measure, then P is Boolean representable.

Thus the argument of Section II--that there are no local representations of P--requires only that the representation should allow a local probability measure.

IV. LOCALITY AND EMBEDDABILITY

According to the analysis of Section II, a local representation of the spin magnitudes of $S + S'$ coincides with a Boolean representation in the sense of Maczyński. Recall that a set Q of magnitudes is Boolean representable if and only if there is a map $\phi: Q \rightarrow C$ from Q into a commutative algebra such that

$$\phi(f(A)) = f(\phi(A))$$

for all $A \in Q$ and any measurable function $f: R \rightarrow R$ (Theorem 4.1). We remark that ϕ is an embedding only if there is a maximal Boolean representation of Q, that is, only if the Boolean representation $(B, \{r_A: A \in Q\})$ has the additional property:

> For every Boolean representation $(B', \{r_A: A \in Q\})$ there is a σ-homomorphism $h: B \rightarrow B'$ such that $h \circ r_A = r'_A$ for every $A \in Q$.

The characterization of a positive solution of the hidden variable problem as the existence of an embedding --in the sense of a one-one homomorphism satisfying Equation 4.1--differs from the characterization in terms of locality by requiring that the Boolean representation be maximal. But this difference is inessential; for by Theorem 4.2 of Maczyński, if Q has a Boolean representation, it has a <u>maximal</u> Boolean representation.

At this point it might be worthwhile to discuss a similar analysis of locality, proposed by Bub (1975). On this view, Bell's analysis of hidden variable theories raises the following question: Can Maczyński's proof that the maximal magnitudes have a classical representation be extended to the maximal and local spin magnitudes of a pair of spin-1/2 particles? The difference between our approach and Bub's is that Bub is somewhat equivocal on the notion of a classical representation. This sometimes means Boolean representation. Certainly this must be intended if there is to be a significant connection with Maczyński's paper. But his discussion of the solution of the specific problem just formulated leaves what is intended by a classical representation less than clear. In outline, Bub's argument that there is no classical representation is roughly this: Let $\alpha, \beta \in B$ be as in Section II with B a product representation of P, that is, α is a subset of B associated with the value of a local spin magnitude referring to S, and β a subset representing an event of S′. For reasons that we need not go into here, Bub argues that the condition that the joint probability of the events represented by α and β is $m_\psi(\alpha \cap \beta)$ implies that their sequential probability is also given by $m_\psi(\alpha \cap \beta)$, and he concludes that there is no classical representation of $K \cup M$ since, with this rule, it is impossible to recover the quantum mechanical sequential probabilities for S + S′. But this argument cannot bear on the question of the Boolean representability of $K \cup M$. Bell and Wigner show--and Bub simply reformulates their argument for sequential probabilities--that for nine maximal magnitudes and six local magnitudes, locality, together with the classical rule for joint probabilities, makes it impossible to recover the statistical relations of ψ. By Theorem 3.1, to exclude a Boolean representation of this particular set of magnitudes it would be necessary to show that the above result holds for every statistical state of the system S + S′. But this is not the case. (See the concluding paragraph of Wigner's paper for a characterization of the class of states for which the argument does hold.)

There is an important methodological consequence of our analysis of locality that has been completely missed in the literature on hidden variables. We have shown that the existence of local hidden variable theories is excluded by the theorem that there is no homomorphism from L into a Boolean algebra B. The correctness of the algebraic analysis of hidden variables depends on the feasibility of requiring the existence of such a homomorphism and therefore rests on the assumption that the algebraic structure of L is physically important. This requirement is generally regarded as physically unmotivated. Now it follows from the principal theorem of this paper that if this requirement is not physically motivated, then neither is the analysis in terms of locality.

But if locality is not physically motivated, then <u>no</u> analysis of the problem hidden variables can be physically motivated.

Only one issue remains open: Someone who accepts the algebraic point of view will apply it to hidden variable theories of single particle systems. Obviously, locality cannot apply here. But locality implies that for many-particle systems, L is physically significant. And it is certainly no great inductive step to the conclusion that L is physically significant even in the case of single particle systems. Hence we may conclude: One who accepts locality, <u>and therefore the algebraic structure of many-particle systems</u>, cannot rationally deny the physical significance of the structure of L in the case of single particle systems.

V. ADDENDA

1. <u>The Boolean Representability of the Local and Maximal Spin Magnitudes</u>[1]

Recall that L is the partial Boolean algebra associated with the ranges of the set P of local and maximal spin magnitudes, that is, $L = \cup\{L_i : i \in P\}$; the elements $a,b \in L$ are compatible in L if there exists an $i \in P$ such that $a,b \in L_i$. L* is the partial Boolean subalgebra of $L(H \otimes H')$ generated by the elements of L. The extension of Theorem 2.1 required by Corollary 2.2 assumes that if a homomorphism is defined on L, it must extend to a homomorphism on L*. The fact that the Boolean subalgebras of L are--while those of L* are <u>not</u>--Boolean representable is a counterexample to this assumption. Hence there can be no purely algebraic argument against a hidden variable interpretation of $S + S'$ that is based solely on locality, that is, that is based solely on the Boolean representability of the set P of local and maximal spin magnitudes.

The proof that P is Boolean representable follows as a corollary to Paul Humphreys' observation that the set I of local and parallel spin magnitudes is Boolean representable. Humphreys represents I by the free σ-product $(B,\{h_{AA}\}_{AA \in M \cap I})$ of the ranges of the parallel magnitudes (i.e., maximal spin magnitudes of the form $A \otimes A$). The range of a local spin magnitude, such as

L_{AI}, is represented by the subalgebra of B generated by $\{h_{AA}(aa_1 \vee aa_2)\}$.

Let L_{AA}, L_{BB} be the ranges of two parallel spin magnitudes, and let $L^*_{AA} = h_{AA}(L_{AA})$, $L^*_{BB} = h_{BB}(L_{BB})$ be their images in B. To define the images L^*_{AB}, L^*_{BA}, of L_{AB}, L_{BA}, let L^*_{AB} be the Boolean subalgebra of B generated by $\{h_{AA}(ai), h_{BB}(ib)$, and let L^*_{BA} be the subalgebra generated by $\{h_{AA}(ia), h_{BB}(bi)\}$. The uniqueness of L^*_{BA} follows, since h_{AA}, h_{IA} are the only maps in $\{h_{IB}, h_{BI}, h_{BB}\}_{IB,BI,BB \in I}$ such that $ai \in \text{dom}(h_{AA})$ and $ai \in \text{dom}(h_{IA})$, and $h_{AA}(ia) = h_{IA}(ia)$ by locality. By the definition of L^*_{BA}, $L^*_{AI} \subset L^*_{AB}$ and $L^*_{IB} \subset L^*_{AB}$. Similarly, $L^*_{IA} \subset L^*_{BA}$ and $L^*_{BI} \subset L^*_{BA}$.

2. Two Extensions of Boolean Representability

Let S be the set of quantum mechanical statistical states for S + S′. Let P′ ⊂ P and let N be the partial Boolean algebra associated with $\{L_i : i \in P'\}$. Let us say that P′ is Boolean representable with respect to S if P′ is Boolean representable and for every statistical state $\alpha \in S$ there exists a probability measure m_α on B such that for every Boolean polynomial $\phi(a_1, \ldots, a_n) \in N$,

$$p_\alpha(\phi(a_1, \ldots, a_n)) = m_\alpha(\phi(h(a_1), \ldots, h(a_n))) .$$

Then Bell and Wigner require that the local and maximal spin magnitudes be Boolean representable in this extended sense. They observed that in the singlet spin state ψ

$$p_\psi(ai \wedge ic) \not\leq p_\psi(ai \wedge ib) + p_\psi(bi \wedge ic)$$

whenever the directions specified by $A \otimes I$, $I \otimes B$, $B \otimes I$, and $I \otimes C$ are coplanar. However,

$$m_\psi(h(ai) \wedge h(ic) \leq m_\psi(h(ai) \wedge h(ib)) + m_\psi(h(bi) \wedge h(ic))$$

for every measure m_ψ on B such that

$$m_\psi(h(bi) \wedge h(ib)) = m_\psi(h(bi)^\perp \wedge h(ib)^\perp) = 0$$

--as required by ψ.

Kochen and Specker's discussion of two spin-1/2 particles in Section 4 of their paper suggests a rather different extension of Boolean representability. Let S and N be as above, and let N* be the partial Boolean subalgebra of $L(H \otimes H')$ generated by N. Call P′ Boolean representable with respect to S and N* if P′ is Boolean representable and for every $\alpha \in S$, there exists an m_α on B such that for every Boolean polynomial $\phi(a_1,\dots,a_n) \in N^*$ in elements $a_1,\dots,a_n \in N$,

$$\text{If } p_\alpha(\phi(a_1,\dots,a_n)) = 1(0) ,$$

$$\text{then } m_\alpha(\phi(h(a_1),\dots,h(a_n)) = 1(0) .$$

Let R denote the requirement of Boolean representability with respect to S, and R* the requirement of Boolean representability with respect to S and N*. Clearly, R* requires preservation of only zero-one probabilities, while R constrains a smaller class of Boolean polynomials; equivalently, R involves a smaller compatibility relation but a larger class of probabilities than does R*.

The first point to notice is that R* is not stronger than R: There exist subsets P′ of P such that R* holds of P′, while R does not hold. For example, to derive the Bell-Wigner contradiction, it suffices that the spin components refer to three coplanar directions of E^3, so that N* may be identical with N in this case.

Recall that a weak embedding between partial Boolean algebras is simply a homomorphism that is one-one on compatible elements. The next three lemmas show that R* is equivalent to the existence of a weak embedding of N* into a Boolean algebra.

LEMMA 1. <u>If the homomorphism</u> h: N -> B <u>extends to a weak embedding</u> h*: N* -> B, <u>then</u> P′ <u>satisfies</u> R*.

<u>Proof</u>. Suppose R* does not hold and let $(a_1,\dots,a_n) \in N^*$ be a Boolean polynomial such that

$$p_\alpha(\phi(a_1,\dots,a_n)) = 1 ,$$

but

$$m(\phi(h(a_1),\ldots,h(a_n)) \neq 1$$

for every $m\colon B \to [0,1]$. Then $m(\phi(h(a_1),\ldots,h(a_n)) = 0$ for every $m\colon B \to \{0,1\}$. Hence, by semisimplicity,

$$h^*(\phi(a_1,\ldots,a_n)) = \phi(h(a_1),\ldots,h(a_n)) = \Lambda_B \,.$$

Since $p_\alpha(\phi(a_1,\ldots,a_n)) = 1$, $\phi(a_1,\ldots,a_n) \neq \Lambda_{N^*}$ and therefore h^* is not a weak embedding. If $p_\alpha(\phi(a_1,\ldots,a_n)) = 0$, $p_\alpha(1 - \phi(a_1,\ldots,a_n)) = 1$ and the argument proceeds as above.

LEMMA 2. _If_ P' _satisfies_ R^*, _then_ N^* _is weakly embeddable into a Boolean algebra_.

Proof. If N^* is not weakly embeddable into a Boolean algebra, there exists a classical tautology $\phi(x_1,\ldots,x_n)$ such that $\phi(x_1,\ldots,x_n)$ is not valid in N^* (cf. Kochen & Specker, 1967, Theorem 4). That is,

$$\phi(a_1,\ldots,a_n) \neq V_{N^*}$$

for some $a_1,\ldots,a_n \in N^*$ in the domain of the propositional function ϕ. Hence,

$$1 - \phi(a_1,\ldots,a_n) \neq \Lambda_{N^*} \,.$$

So there exists a state $\alpha \in S$ such that

$$p_\alpha(1 - \phi(a_1,\ldots,a_n)) = 1 \,.$$

(The set of states of N^* is _sufficient_; cf., e.g., Axiom E4 of Zierler, 1961.) By the definition of N^*,

$$a_i = \alpha_i(a_{i1},\ldots,a_{ik(i)}) \,,$$

where $a_{i1},\ldots,a_{ik(i)} \in N$. Thus,

$$p_\alpha(1 - \phi(\alpha_1,\ldots,\alpha_n)) = 1 \,.$$

But for no $m\colon B \to [0,1]$ do we have

$$m(1 - \phi(h(\alpha_1),\ldots,h(\alpha_n))) = 1 \,,$$

since $1 - \phi(x_1,\ldots,x_n)$ is a classical contradiction.

For the next lemma, we assume that B is the Boolean algebra of a <u>maximal</u> Boolean representation of P′. By Theorem 4.2 of Maczyński, such a representation exists if a Boolean representation exists. (It is in fact the direct limit of the ranges of the magnitudes in P′.) Hence, this introduces no restriction in generality.

LEMMA 3. <u>If N* is weakly embeddable into a Boolean algebra, the homomorphism h: N -> B from N into the Boolean algebra representing P′ extends to a weak embedding of N* into B.</u>

<u>Proof</u>. Let f: N* -> C be a weak embedding of N* into C. For each $i \in P'$, $f_i = f|L_i$ is a monomorphism from L_i into C, and $(C,\{f_i: i \in P'\})$ is a Boolean representation of P′ (cf. Maczyński, Theorem 4.1). Since $(B,\{h_i: i \in P'\})$ is maximal, there is a homomorphism k: B -> C such that for every $i \in P'$, $f_i = k \circ h_i$. Hence, each h_i is a monomorphism and h: N -> B is therefore a weak embedding. Let C′ be the subalgebra of C generated by f[N*]. Since N generated N*, f[N] generates C′, and therefore $C' \cong B$.

The preceding three lemmas together imply the following theorem.

THEOREM. <u>P′ satisfies R* if and only if N* is weakly embeddable into a Boolean algebra</u>.

Is R stronger than R*? This would be the case if it could be shown that any subset P′ of P for which R* does not hold allows the derivation of the Bell-Wigner contradiction. This is true, for example, of the partial subalgebra constructed by Kochen and Specker, since it contains the spin components of three coplanar directions in E^3 (cf. proof of Lemma 1 and the embedding described in Sec. 4). Hence, in this case, R does not hold. I see no reason to suppose that this case is atypical, but I do not have a proof. If this could be shown, then although the algebraic component of R is strictly weaker than that of R*, the total set of probabilistic and algebraic assumptions incorporated in R is stronger than the total set of probabilistic and algebraic assumptions of R*.

The methodological claim that the Bell-Wigner approach actually supports the algebraic analysis of the hidden variable question would then be sustained.

3. Local Theories of S + S′

In this paper, a theory of the system S + S′ is regarded as local if the partial Boolean algebra on which the local and maximal spin magnitudes are based preserves the functional relations holding in P. When this partial Boolean algebra is required to be a Boolean algebra, this condition is precisely what is formalized by the notion of a Boolean representation. It is clear that there are three main local theories of S + S′.

There is first the local theory that is based on B--the direct limit of the family $\{L_i : i \in P\}$. Second, the theory based on $L(H \otimes H')$--or the relevant partial Boolean subalgebra, L*--preserves the functional relations of P and is therefore local. Of these two theories, only the non-Boolean representation by subspaces of Hilbert space is consistent with the statistical predictions of quantum mechanics.

A third possibility is the theory based on the partial Boolean algebra L. Since L is homomorphic to a Boolean algebra, this theory admits an extension to one that contains two-valued measures and, hence, dispersion-free states. But since L is not a Boolean algebra, it is also clear that the states of this theory--which may be identified with the restrictions of the quantum states to L--are not constrained by Bell′s inequality. Moreover, there is a natural sense in which such a theory is classical; namely, the set of its statistical states admits an extension to one that includes a generous supply of two-valued measures. Thus, if the problem is merely to represent the functional relations of P and the statistical predictions of S by a theory that is classical in this sense, then it would seem that this can be accomplished by a theory of S + S′ based on L. In fact, L is a σ-class in the sense of Gudder (1976). Hence, such a representation of S + S′ is a genuine theoretical possibility. As it stands, the Bell-Wigner analysis can exclude such a resolution only if it can support the demand that every classical theory of S + S′ must be based on a Boolean algebra.

On the algebraic approach, the situation is rather different. Since R* is equivalent to the existence of a weak embedding, a classical extension must contain enough two-valued measures to separate every nonnull element of L* from the zero of L*; by the analysis of Section 4 of Kochen and Specker, this condition must fail for the triplet spin subspace. Notice that although the classical extension must contain two-valued measures, there is nothing in the algebraic approach that requires that the classical theory of S + S′ should be based on a Boolean algebra. The difficulty here is to motivate the demand that a theory of S + S′ be based on L*. Certainly, the principal physical reason is that L* contains the basis for a representation of a system of spin s = 1, so that a theory of S + S′ based on L* expresses a fundamental discovery of the quantum theory of spin, namely, the fact that two spin-1/2 particles may combine to form a single 'particle′ of spin s = 1 or spin s = 0 (cf. Schiff, 1955, Sec. 41). A theory of S + S′ based solely on L cannot express this possibility. For this reason, I believe such a theory should be viewed as an intermediate step to one based on L* rather than an alternative to the quantum mechanical representation of two spin-1/2 particles.

References

Bell, J. S. On the Einstein-Podolsky-Rosen paradox. Physics, 1964, 1, 195-200.

Bell, J. S. On the problem of hidden variables in quantum mechanics. Reviews of Modern Physics, 1966, 38, 447-452.

Bub, J. Randomness and locality in quantum mechanics. In P. Suppes (Ed.), Logic and probability in quantum mechanics. Dordrecht: Reidel, 1975.

Demopoulos, W. Remark on a paper of Maczyński. Reports on Mathematical Physics, 1976, 9, 171-176.

Dwinger, P. Direct limits of partially ordered systems of Boolean algebras. Indagationes Mathematicae, 1967, 29, 317-325.

Fine, A. On the completeness of quantum theory. In P. Suppes (Ed.), Logic and probability in quantum mechanics. Dordrecht: Reidel, 1975.

Gudder, S. On hidden variable theories. Journal of Mathematical Physics, 1970, 11, 431-436.

Gudder, S. A generalized measure and probability theory for the physical sciences. In W. L. Harper & C. A.

Hooker (Eds.), Foundations of probability theory, statistical inference and statistical theories of science (Vol. 3). Dordrecht, The Netherlands: Reidel, 1976.

Kochen, S., & Specker, E. P. The problem of hidden variables in quantum mechanics. Journal of Mathematics and Mechanics, 1967, 17, 59-87.

Maczyński, M. J. Boolean properties of observables in axiomatic quantum mechanics. Reports on Mathematical Physics, 1971, 2, 135-150.

Schiff, L. Quantum mechanics (3rd ed.). New York: McGraw-Hill, 1955.

Sikorski, R. Boolean algebras. Berlin: Springer Verlag, 1964.

Suppes, P., & Zanotti, M. On the determinism of hidden variable theories with strict correlation and conditional statistical independence of observables. In P. Suppes (Ed.), Logic and probability in quantum mechanics. Dordrecht: Reidel, 1975.

Wigner, E. On hidden variables and quantum mechanical probabilities. American Journal of Physics, 1970, 38, 1005-1009.

Zierler, N. Axioms for non-relativistic quantum mechanics. Pacific Journal of Mathematics, 1961, 11, 1151-1169.

Notes

Research for this paper was partially supported by Canada Council Grant S74 1032. I wish to thank Allen Stairs and Vandana Shiva for many rewarding discussions on the subjects dealt with in this paper.

[1]The possibility of constructing a Boolean representation of the local and maximal spin magnitudes was first raised by Allen Stairs.

A NOTE ON DEMOPOULOS'S PAPER "LOCALITY AND THE ALGEBRAIC STRUCTURE OF QUANTUM MECHANICS"

Paul W. Humphreys
Stanford University*

Demopoulos has indicated in the preceding paper that there is no Boolean representation of the class P of all locally maximal and maximal magnitudes associated with a system $S + S'$ (represented by a Hilbert space $H_2 \otimes H_2'$).[1] I shall show here that a particular subset of that set of magnitudes does have a Boolean representation. The particular form of this subset will indicate that Demopoulos's argument for his strong locality condition is necessary, and that it is the presence of functional relations between incompatible magnitudes that blocks Boolean representation.

For simplicity in what follows, I shall restrict considerations to the case of a system of two spin-1/2 particles that are spatially separated. Suppose we have a local spin magnitude $A \otimes I$, and two maximal spin magnitudes $A \otimes B$ and $A \otimes C$, such that $A \otimes I = f(A \otimes B)$ and $A \otimes I = g(A \otimes C)$. Within the axiomatic tradition of Mackey, this has been taken to mean $P(A \otimes I, \psi, E) = P(A \otimes B, \psi, f^{-1}(E))$ for all $E \in B(R^2)$ and all statistical states ψ. Similarly for $A \otimes C$. In order to argue that a deterministic representation of the magnitudes should be a Boolean representation, Demopoulos extends this identity condition to a pointwise identity holding between the values of $A \otimes I$ and $A \otimes B$, at least in the case of the separated systems we are considering. Bell (1966) and Gudder (1970) have previously voiced objections against such pointwise identification of functions (in the case of isolated systems), on the grounds that if $A = f(B)$ and $A = g(C)$, where B,C are incompatible, the possibility of interference between measurements of B and C precludes this type of identity

---

*Presently at the University of Virginia.

Studies in the Foundations of Quantum Mechanics, pp. 145-147

and makes only statistical identification of operators possible.

Although Demopoulos has demonstrated that there are no 'local' representations of P using a probabilistic argument, the interesting case seems to me to be the claim that locality implies Boolean representability, and the resultant argument against a deterministic representation of the observables. This first argument avoids probabilistic concepts altogether, which, for those of us viewing quantum mechanics as an essentially probabilistic theory, is a striking achievement. I thus see the counterarguments against Bell and Gudder as a central feature of the paper, and I should like to indicate why those counterarguments are indispensable for Demopoulos.

Let us call those maximal spin magnitudes $A \otimes A$ on $H_2 \otimes H_2'$ where the spin measurement is made in the same direction on both S and S′ the <u>parallel</u> magnitudes. If it were possible to restrict ourselves to the class of all locally maximal and parallel magnitudes, and prove that this class was not Boolean representable, then the objections of Bell and Gudder could be avoided completely. However, it turns out that this restricted collection of magnitudes is Boolean representable.

THEOREM. <u>The set of all parallel and local magnitudes</u> P <u>is Boolean representable</u>.

<u>Proof</u>. The partial ordering on the set $\{L_{AI}, L_{IA}, L_{AA}\}$ is determined by $L_{AI} \leq L_{AA}$, $L_{IA} \leq L_{AA}$, and the fact that each subfamily $\{L_{IA}\}$, $\{L_{AI}\}$, $\{L_{AA}\}$ is totally unordered with respect to the partial ordering $\leq$: i.e., $L_{AI} \leq L_{BI}$ implies $L_{AI} = L_{BI}$. We show that the direct limit of $\{L_{AI}\} \cup \{L_{IA}\} \cup \{L_{AA}\}$ is nondegenerate. This limit is in fact the free σ-product of the elements of $\{L_{AA}\}$. Let the members of the partial ordering be indexed by $i,j \in I$, where $i \leq j$ if $L_i \leq L_j$ and let $\emptyset_{ij}: L_i \rightarrow L_j$ be the inclusion homomorphism. Let $\{h_j\}$ be the monomorphisms $h_j: L_j \rightarrow B$, where $L_j \in \{L_{AA}\}$ and B is the free σ-product of the members of $\{L_{AA}\}$. Then define the homomorphisms h_i for $i \leq j$ by $h_i = h_j \circ \emptyset_{ij}$.

Then $(B,\{h_k\}_{k\in I})$ is the direct limit. For $h_j \circ \emptyset_{ij} = h_i$ by definition and the existence of the unique homomorphism associating $(B,\{h_k\}_{k\in I})$ with any other Boolean representation $(C,\{g_k\}_{k\in I})$ is guaranteed by a necessary condition on free σ-products (see Sikorski, 1964, p. 178, Theorem 38.3). Furthermore, B is nondegenerate. Thus, using Theorem 4.2 of Maczyński (1971), P is Boolean representable.

This result demonstrates two connected aspects of the situation. First, there are functional relations between the $A \otimes I$ and the $A \otimes A$, and these are preserved in the representation. These functional dependencies are all of the form $A \otimes I = f(A \otimes A)$, and hence it is the presence of functional connections between incompatible magnitudes that prevents the embedding into a Boolean algebra when we consider the full set of magnitudes. This, of course, is the essential feature of all recent algebraic proofs against hidden variables. The second feature has already been mentioned--that the locality assertion allows us to disregard any interaction of incompatible magnitudes in a distant system.

Note

[1]For the definition of a Boolean representation, see Demopoulos's paper. The technical apparatus used here may be found either in that article or in the paper by Maczyński (1971). See also Bub's comments in this volume concerning the proof.

I should like to thank Demopoulos for a correction to an earlier version of this paper.

References

Bell, J. S. On the problem of hidden variables in quantum mechanics. *Reviews of Modern Physics*, 1966, *38*, 447-452.

Gudder, S. On hidden variable theories. *Journal of Mathematical Physics*, 1970, *11*, 431-436.

Maczyński, M. J. Boolean properties of observables in axiomatic quantum mechanics. *Reports on Mathematical Physics*, 1971, *2*, 135-150.

Sikorski, R. *Boolean algebras*. Berlin: Springer-Verlag, 1964.

COMMENT ON "LOCALITY AND THE ALGEBRAIC STRUCTURE OF QUANTUM MECHANICS," BY WILLIAM DEMOPOULOS

Jeffrey Bub
University of Western Ontario

The Demopoulos argument, that Bell's locality condition implies the existence of a homomorphism from L(H) into a Boolean algebra, and hence into the 2-element Boolean algebra, is fallacious.

Expressed in the language of the Demopoulos paper, the locality condition for the Hilbert space $H_n \otimes H'_m$ is this: Consider any local proposition x associated with the system S. Let L_x be the local Boolean subalgebra of S generated by x, i.e., $(0,x,x^{\perp},1)$. For any pair of extensions, L_j and L_k, of L_x to maximal Boolean subalgebras in $L(H_n \otimes H'_m)$, <u>such that L_j and L_k differ only with respect to local elements associated with</u> S′ (i.e., elements belonging to local subalgebras of S′ but not to local subalgebras of S), we require $h_j(x) = h_k(x)$ (where h_j and h_k are the homomorphisms of the Boolean representation, i.e., homomorphisms of L_j and L_k into a Boolean algebra--see the Demopoulos proof of Theorem 2.1).

In the 2×2 dimensional case--the only case explicitly considered--the argument depends on the assumption that a Boolean representation of the local and maximal magnitudes exists if and only if the locality condition is satisfied. It is then argued that the existence of a Boolean representation for the local and maximal spin magnitudes of S + S′ implies the existence of a homomorphism from $L(H_2 \otimes H'_2)$ into a Boolean algebra (Theorem 2.1).

Since the result purports to be quite general, it is instructive to consider how this argument might be extended to the 3×3 dimensional case, or the $n \times m$ dimensional case in general. Here the locality condition is clearly much weaker than the condition of Boolean

Studies in the Foundations of Quantum Mechanics, pp. 149-153

representability for the local and maximal magnitudes. The obvious generalization of Theorem 2.1 would involve taking locality as equivalent to the Boolean representability of the maximal and locally maximal[1] magnitudes and then showing that if the set of maximal and locally maximal magnitudes on $H_n \otimes H'_m$ is Boolean representable, there exists a homomorphism from $L(H_n \otimes H'_m)$ into a Boolean algebra.[2] This approach fails, as can easily be seen by the following example, which also suggests a counterexample in the 2×2 dimensional case.

Suppose the maximal Boolean subalgebra L_j in the extension of Theorem 2.1 to the 3×3 dimensional case is generated by a triple of orthogonal vectors a_1, a_2, a_3 in H_3 and a'_1, a'_2, a'_3 in H'_3. Let L_k be generated by the triple b_1, b_2, b_3 in H_3 and a'_1, a'_2, a'_3 in H'_3, where b_1 and b_2 are rotated relative to a_1, a_2 in the a_1, a_2 plane and $b_3 = a_3$.

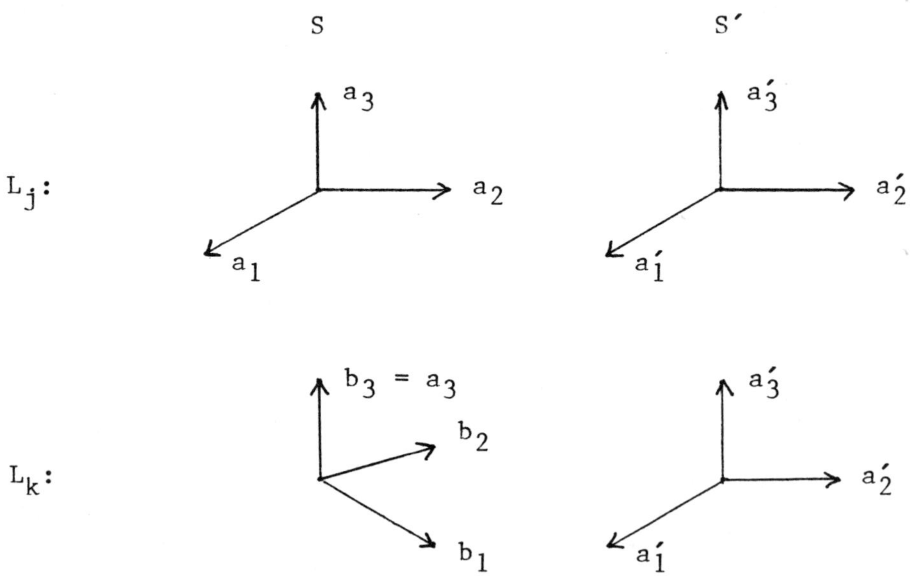

Consider the element corresponding to a_3 in $L_j \cap L_k$ (i.e., the element in $L(H_3 \otimes H'_3)$ corresponding to the projection operator $P_{a_3} \otimes I'$). Call this element x. Now, x belongs to the locally maximal Boolean subalgebra L_a generated by a_1, a_2, a_3, and also to the locally maximal

Boolean subalgebra L_b generated by b_1, b_2, b_3. It also belongs to the local (but not locally maximal) subalgebra L_x generated by x, i.e., $(0, x, x^{\perp}, 1)$. The proof of the theorem requires that $h_j(x) = h_k(x)$. But this is not required by the locality condition, for L_j and L_k are not extensions of L_x that differ only with respect to local elements associated with S′--i.e., we might have the proposition $x \in L_j$ true and $x \in L_k$ false.

By contrast, if L_j and L_k are generated by the vectors

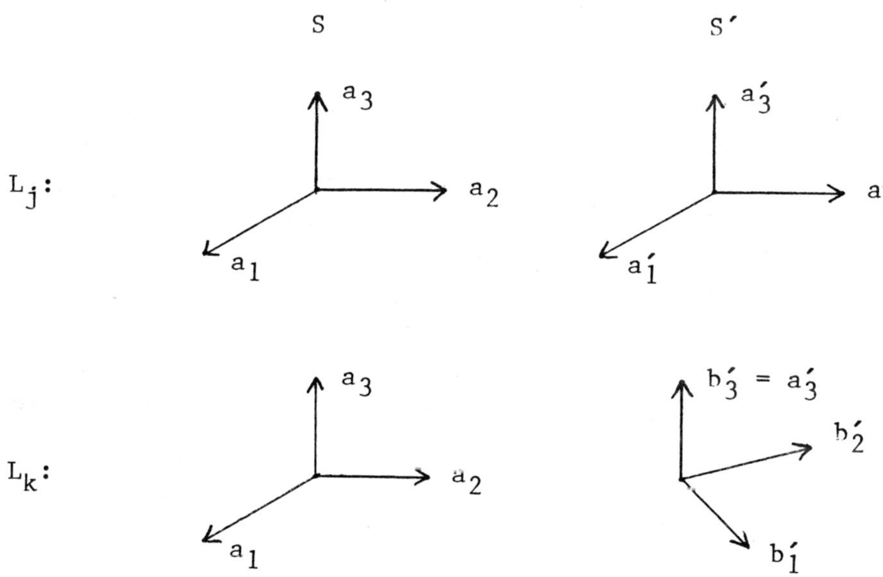

then locality does require that $h_j(x) = h_k(x)$, where x is again the element corresponding to a_3 in $L_j \cap L_k$.

The crucial point in the construction of the homomorphism in Theorem 2.1 or its extension to the $n \times m$ dimensional case is the claim that $h_j(x) = h_k(x)$, that is, that the values of the homomorphisms h_j, h_k corresponding to two maximal magnitudes coincide on subspaces of L belonging to the union of the ranges of the locally maximal magnitudes. I have pointed out that the locality condition only requires this equality to hold for certain L_j, L_k (depending on the element x). Similarly, the Boolean representability of the maximal and locally maximal magnitudes only requires that, for $x \in L_{AI}$, say, where L_{AI} is locally maximal, $h_j(x) = h_k(x)$ for any pair of maximal Boolean subalgebras L_j and L_k *which are maximal extensions of* L_{AI}. By requiring that $h_j(x) = h_k(x)$ for every maximal L_j, L_k containing x, the argument at most reduces to a proof of the theorem that Boolean representability of the *maximal and local* magnitudes on $H_n \otimes H'_m$ implies the existence of a homomorphism from $L(H_n \otimes H'_m)$ into a Boolean algebra. And this theorem evidently relates the existence of a homomorphism to a far stronger condition than locality. The following counterexample in the 2×2 case, where all local magnitudes are locally maximal, shows that the argument does not prove even *this* theorem.

At first sight, it appears that the argument goes through in the 2×2 dimensional case, because here we can have $x \in L_j$, $x \in L_k$, $L_j \neq L_k$, only if L_j and L_k are generated by

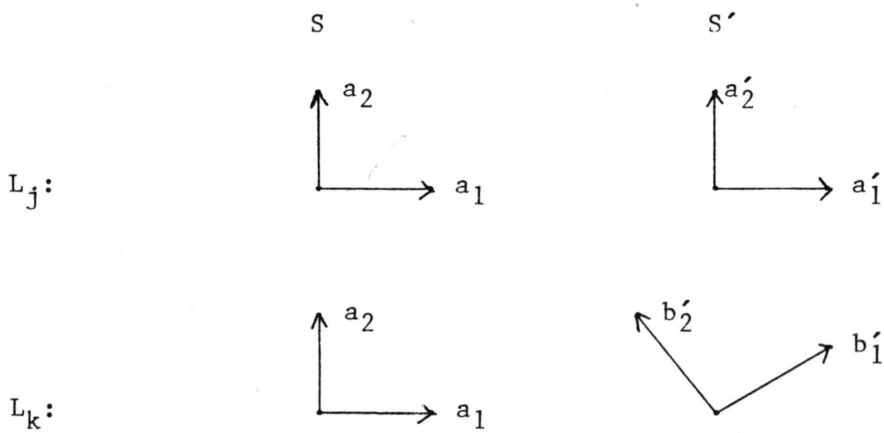

where x is the element corresponding to a_2 in $L_j \cap L_k$. But then locality *does* require that $h_j(x) = h_k(x)$.

Now, consider any general vector v in $H_2 \otimes H'_2$ that is not of the form $a \otimes b'$, that is, any linear combination of such vectors (e.g., the singlet spin state itself: $1/\sqrt{2}\, a_1 \otimes a'_2 - 1/\sqrt{2}\, a_2 \otimes a'_1$). Let L_j be the maximal

Boolean subalgebra in $L(H_2 \otimes H_2')$ generated by v and any triple of vectors orthogonal to v, and let L_k be the maximal Boolean subalgebra generated by v and any other triple of vectors orthogonal to v. Call x the element in $L(H_2 \otimes H_2')$ corresponding to v. Then certainly locality does not require that $h_j(x) = h_k(x)$, and neither does the Boolean representability of the maximal and local magnitudes. Note that x belongs to the maximal Boolean subalgebras L_j and L_k, and also to the nonmaximal--but nonlocal!--Boolean subalgebra $(0, x, x^{\perp}, 1)$, as well as to a continuous infinity of other nonmaximal and nonlocal Boolean subalgebras. Nothing in the Demopoulos argument guarantees that the map h defined on the elements in the local and maximal Boolean subalgebras of the type L_{AI}, L_{IB}, L_{AB} can be extended to a homomorphism on all of $L(H_2 \otimes H_2')$, which contains elements like x. Putting this another way: Not all maximal Boolean subalgebras of $L(H_2 \otimes H_2')$ are of the type L_{AB}, and not all nonmaximal Boolean subalgebras are local subalgebras, i.e., of the type L_{AI} or L_{IB}. (In general, the restriction to subalgebras of the type L_{AB}, L_{AI}, L_{IB} will exclude subalgebras generated by any magnitude which is additive on S_1 and S_2, e.g., the total energy, even if the systems are noninteracting.)

Notes

Discussions with Allen Stairs are gratefully acknowledged.

[1] A locally maximal magnitude of $S + S'$ is a nonmaximal magnitude of the form $A \otimes I'$ (or $I \otimes A'$), where A is maximal in H_n (or A' is maximal in H_m').

[2] Such a generalization was included in an Appendix to the original version of Demopoulos's paper, which seems to have suffered an Appendectomy.

ON THE POSSIBILITY OF SPONTANEOUS PROJECTION POSTULATES

Stephen Norman
Stanford University

I. INTRODUCTION

It is usually assumed in conventional quantum mechanics that the state of a system after measurement is the mixture given by the von Neumann projection postulate:

$$U' = \sum_i (Ua_i, a_i) P_{[a_i]}$$

where U is the statistical operator describing the state of the system immediately before measurement, and the a_i form a complete orthonormal set drawn from the eigenstates of the operator which corresponds to the observable being measured. Dissatisfaction with the postulate has been generated by each of the following facts: First, the change or reduction that the postulate describes is discontinuous. Second, the reduction is `alocal' in the sense that the reduction takes place without regard for spatial extension. Third, the postulate gives a special status to measurement by proposing that systems undergo a special kind of transition when measured, which is quite different from the normal evolution of the system. Von Neumann (1955) put this latter objection very aptly: "Indeed, a physical intervention can be nothing else than the temporary insertion of a certain energy coupling into the observed system. . . . Why then do we need the special process 1. for the measurement?" (p. 352).

One might try to overcome this last difficulty (but neither of the first two) by retaining reductions of the kind described by the von Neumann postulate but severing the connection between such reductions and measurement. That is, one might suppose that such reductions take place spontaneously and quite independently of whatever measurements are performed. Of course, a move like this

Studies in the Foundations of Quantum Mechanics, pp. 155-164

will, in general, lend philosophical respectability at the cost of empirical adequacy. Von Neumann (1955, p. 212) was, after all, constrained to introduce his projection postulate by consideration of the experimental results of Compton and Simons.

Nevertheless, Nancy Cartwright (1978, p. 54; 1979), noting some special properties of exponentially decaying systems, has argued in favor of just such a move, coupled with some restrictions on the class of observables. The necessity of taking this latter step becomes immediately apparent when one tries to construct spontaneous projection postulates whose observational consequences will be indistinguishable from those of the conventional projection postulate. Let me emphasize here that Cartwright does not advocate changing the form of reductions, but only their occasion. The idea is that quantum systems evolve through time in accordance with the Schrödinger equation and unreduced by measurement but sometimes reduce spontaneously and instantaneously in the manner prescribed by the von Neumann projection postulate.

Strictly speaking, the von Neumann projection postulate applies in conventional quantum mechanics only in the case of certain ideal measurement setups. In other cases, for example, the measurement may destroy the measured system. In those cases where conventionally the von Neumann projection postulate is appropriate, the Cartwright system, in the absence of any reduction, will be unchanged by the measurement. In the remainder of this paper we shall be considering only these ideal measurements, which do not disturb the Cartwright system.

In the next section of this paper, I will set out the objectives, as I understand them, of Cartwright's program. This will be followed, in Section III, by an examination of exponential decay and its significance for spontaneous projection postulates. Finally, in Section IV, I will prove a theorem which shows that no theory of spontaneous reduction can have, in a certain ideal but precise sense of the phrase, the same empirical content as conventional quantum mechanics, unless the only observables permitted are all constants of the motion. But in that case, according to the theory, no system undergoes observable change, which is obviously an unsatisfactory state of affairs. It is a minor but interesting corollary of the theorem that exact exponential decay can never occur in quantum mechanics.

II. A SUMMARY OF CARTWRIGHT'S PROPOSAL

Here, and in the rest of this paper, the symbol $U_s(t)$ will be used to denote the state of a system at time t according to a spontaneous projection postulate, and $U_q(t)$ will be used for the state according to conventional quantum mechanics, where the particular feature that we wish to pick out here by use of the phrase "conventional quantum mechanics" is reduction, in accordance with the von Neumann projection postulate, <u>upon measurement</u>.

The objectives of Cartwright's program (as I understand them to be, from conversations with her and from Cartwright [1979]) are:

(1) to produce a spontaneous projection postulate which specifies, for each initial state of each quantum system, the times at which and the eigenstates over which that system can subsequently reduce spontaneously, and to produce a set of <u>allowed observables</u>, such that

(2) the (statistical) state $U_s(t)$ of the system for all $t > t'$ is determined by the spontaneous projection postulate (and, of course, the Schrödinger equation, which determines the evolution of the system in between the spontaneous reductions) when $U_s(t')$ is known, and

(3) at any time t and for any possible value a of any allowed observable A

$$(U_s(t)\alpha, \alpha) = (U_q(t)\alpha, \alpha)$$

where $U_s(t)$ and $U_q(t)$ are as defined above, and α is the eigenfunction corresponding to a.

The third condition is intended to give precise expression to the notion of "having the same empirical content as," and it can be seen immediately that satisfying the condition will not be easy, for $U_s(t)$ is unaffected by ideal measurements made on the system prior to t, whereas $U_q(t)$, in general, is strongly affected. The reason for this is that an ideal measurement at t' causes $U_q(t')$ to reduce in accordance with the von Neumann projection postulate, whereas $U_s(t')$ is unchanged by the measurement. It can also be seen why some restrictions on observables will certainly be necessary in order to produce a postulate which satisfies this third condition. For if one allows each Hermitian operator to represent an observable, then, when $U_q(t)$ is different from $U_s(t)$, as often it must be, there will in general be some measurement which reveals this fact. We shall say of a spontaneous projection postulate that satisfies this third condition that it "preserves the quantum statistics."

III. SPONTANEOUS REDUCTION: AN EXEMPLAR

In order to illustrate the special properties of exponential decay vis-à-vis spontaneous projection postulates, we imagine a simple quantum system Σ with the following characteristics: D is an observable of Σ whose operator $\hat{D}$ has two eigenfunctions u and d with eigenvalues 1 and 0, respectively. The Hamiltonian H of Σ is such that

$$\exp(-iHt)u = [\exp(-kt)]^{1/2}u + [1 - \exp(-kt)]^{1/2}d ,$$

$$\exp(-iHt)d = d .$$

In other words, the system decays exponentially out of u into d, but once in d, it stays there.

It is convenient to introduce some notation at this point. We let $U_\sigma(t)$, where σ is a sequence $\langle t_0, t_1, \ldots, t_n, t_f \rangle$ and $t_0 \leq t_1 < t_2 < \ldots < t_n \leq t_f$, denote the (statistical) state of Σ at t given that in the interval from t_0 up to t_f the system reduces over u and d at times $t_1, t_2, \ldots, t_n$ (and only at those times). These reductions, of course, are of the form given by the von Neumann projection postulate

$$U \rightarrow (Uu,u)P_{[u]} + (Ud,d)P_{[d]} .$$

We now prove the following theorem.

THEOREM. <u>If</u>

$$U_\sigma(t_0) = U_\tau(t_0) = \lambda P_{[u]} + (1 - \lambda)P_{[d]}, \quad 0 \leq \lambda \leq 1,$$

<u>then</u>

$$(U_\sigma(t_f)u,u) = (U_\tau(t_f)u,u)$$

<u>and</u>

$$(U_\sigma(t_f)d,d) = (U_\tau(t_f)d,d)$$

<u>for all</u> $t_0, t_1, \ldots, t_n, t_f$ <u>where</u> $\sigma = \langle t_0, t_1, \ldots, t_n, t_f \rangle$ <u>and</u> $\tau = \langle t_0, t_f \rangle$.

Put informally, this theorem says that if Σ is initially in a mixture of u and d, then measurement of D at a later time will yield D = 0,1 with the same probabilities no matter what reductions over u and d have taken place in the meantime.

The proof is by induction on n.

<u>Basis case</u>: n = 0. In this case no reductions at all take place between (and including) t_0 and t_f, so $\sigma = \tau = \langle t_0, t_f \rangle$ and the theorem holds identically.

<u>Induction</u>. Suppose that the theorem holds for all $t_0, t_1, \ldots, t_m, t_f$. We need to show that it holds for all $t_0, t_1, \ldots, t_{m+1}, t_f$. Let $\rho = \langle t_0, t_1, \ldots, t_{m+1}, t_f \rangle$ and $\sigma = \langle t_0, t_1, \ldots, t_m, t_f \rangle$. Now we know $U_\sigma(t_m) = \mu P_{[u]} + (1 - \mu)P_{[d]}$ for some μ, $0 \leq \mu \leq 1$, because a reduction takes place over u and d at $t = t_m$. So $U_\sigma(t_f) = \mu P_{[\exp(-iH(t_f-t_m))u]} + (1 - \mu)P_{[\exp(-iH(t_f-t_m))d]}$, remembering that according to σ no reduction takes place between t_m and t_f; hence,

$$(U_\sigma(t_f)u,u) = \mu \exp(-k(t_f - t_m))$$

and

$$(U_\sigma(t_f)d,d) = 1 - \mu \exp(-k(t_f - t_m)) .$$

On the other hand, if an additional reduction takes place at time t_{m+1},

$$U_\rho(t_{m+1}) = \mu \exp(-k(t_{m+1} - t_m))P_{[u]} + [1 - \mu \exp(-k(t_{m+1} - t_m))]P_{[d]} ,$$

$$U_\rho(t_f) = \mu \exp(-k(t_{m+1} - t_m))P_{[\exp(-iH(t_f-t_{m+1}))u]} + [1 - \mu \exp(-k(t_{m+1} - t_m))] \cdot P_{[\exp(-iH(t_f-t_{m+1}))d]} .$$

So

$$(U_\rho(t_f)u,u) = \mu \exp(-k(t_{m+1} - t_m))\exp(-k(t_f - t_{m+1}))$$
$$= \mu \exp(-k(t_f - t_m)) .$$

Similarly,

$$(U_\rho(t_f)d,d) = 1 - \mu \exp(-k(t_f - t_m)) .$$

Thus $(U_\rho(t_f)u,u) = (U_\sigma(t_f)u,u)$,

$(U_\rho(t_f)d,d) = (U_\sigma(t_f)d,d)$;

but $(U_\sigma(t_f)u,u) = (U_\tau(t_f)u,u)$

and $(U_\sigma(t_f)d,d) = (U_\tau(t_f)d,d)$, by inductive hypothesis.

So $(U_\rho(t_f)u,u) = (U_\tau(t_f)u,u)$

and $(U_\rho(t_f)d,d) = (U_\tau(t_f)d,d)$,

which completes the inductive step and establishes the theorem.

Naturally, this indifference of the statistics to prior reduction is a consequence of the special properties of exponential decay and does not occur in general. However, exponential decay, or almost exponential decay, is not at all an exceptional condition in quantum mechanics if one picks the right basis (u and d in this case). It can be shown (see, e.g., Goldberger & Watson, 1964) that a system in an eigenstate of an operator H_0 (with total Hamiltonian $H_0 + V$) decays almost exponentially and irreversibly out of that eigenstate if the total Hamiltonian has a continuous spectrum and the perturbation V is small.

In the case of Σ, we can use the indifference of the statistics to prior reduction to obtain a fragment of a spontaneous projection postulate which preserves the quantum statistics. A suitable fragment would be one which specified that the only observable was D, and that, whenever the system is in a state

$$U_s(t) = \mu P_{[u]} + (1 - \mu)P_{[d]} ,$$

the next reduction takes place over u and d at time

$t + \Delta t$, where Δt is some constant. If the system is prepared so that initially

$$U_s(t_0) = U_q(t_0) = \lambda P_{[u]} + (1 - \lambda)P_{[d]} ,$$

then by the theorem we have just proved, it immediately follows that, whatever ideal D measurements are performed on the system,

$$(U_s(t')u,u) = U_q(t')u,u) ,$$

$$(U_s(t')d,d) = U_q(t')d,d) , \quad \text{for all } t' \geq t.$$

Thus, the third condition of Section II is satisfied and we have preservation of the quantum statistics, in these particular circumstances. We shall not discuss the limitations of this fragment, nor ways of extending it, because in the next section we shall prove a theorem which undercuts the whole enterprise of constructing spontaneous projection postulates which preserve the quantum statistics.

IV. THE IMPOSSIBILITY OF PRESERVING THE QUANTUM STATISTICS

The discussion of the previous section notwithstanding, we now prove the following.

THEOREM. _If a spontaneous projection postulate preserves the quantum statistics, the observables which it allows must all be constants of the motion._

Proof. Let A be any allowed observable, let $\{\alpha_i\}$ be a complete orthonormal set of eigenvalues of $\hat{A}$ such that, according to conventional quantum mechanics, the system undergoes the reduction

$$U_q(t) \rightarrow \sum_i (U_q(t)\alpha_i,\alpha_i)P_{[\alpha_i]}$$

upon ideal measurement of A at t. By hypothesis, since the spontaneous projection postulate preserves the quantum statistics,

$$(1) \qquad (U_s(t)\alpha_k,\alpha_k) = (U_q(t)\alpha_k,\alpha_k)$$

for all k and all times t (by <u>all times</u> we mean, of course, all times during the lifetime of the system). Let an ideal A measurement be performed at $t = t_0$ and suppose that no other measurements are performed on the system. At $t = t_0$, then, according to conventional quantum mechanics, the system undergoes the reduction

$$U_q(t_0) \to \sum_i (U_q(t_0)\alpha_i,\alpha_i) P_{[\alpha_i]}$$

and the state of the system at $t \geq t_0$ is given by

$$(2) \qquad U_q(t) = \sum_i (U_q(t_0)\alpha_i,\alpha_i) P_{[\exp(-iH(t-t_0))\alpha_i]} \,.$$

From (1) and (2) we obtain

(3)

$$(U_s(t)\alpha_k,\alpha_k) = \sum_i (U_q(t_0)\alpha_i,\alpha_i) |(\exp(-iH(t - t_0))\alpha_i,\alpha_k)|^2 .$$

Differentiating (3) with respect to t_0, setting t_0 to t, and using the orthonormality of the α_i, we obtain

$$(4) \qquad 0 = \frac{\partial}{\partial t_0} (U_q(t_0)\alpha_k,\alpha_k) \Big|_{t_0 = t} + (U_q(t)\alpha_k,\alpha_k)(i(H\alpha_k,\alpha_k) - i(\alpha_k,H\alpha_k)) \,.$$

The reason that the left-hand side of (3) vanishes upon differentiation with respect to t_0 is that ideal measurement of A at t_0 does <u>not</u>, according to the spontaneous projection postulate, change the state of the system and therefore $(U_s(t)\alpha_k,\alpha_k)$ is actually independent of t_0. However, H, which represents the total energy of the system, must be Hermitian in order that the energy have real expectation value, in which case the second term of (4) vanishes and so

$$\frac{\partial}{\partial t_0} (U_q(t_0)\alpha_k,\alpha_k) \Big|_{t_0 = t} = 0 \quad \text{for arbitrary } t$$

and hence

$$(U_q(t)\alpha_k,\alpha_k) = c_k ,$$

(where c_k is a time-independent constant) for all k. So A is indeed a constant of the motion.

The critical occurrence in this proof is the vanishing of the second term of (4). This term, when expressed in its original form

$$\sum_i (U_q(t_0)\alpha_i,\alpha_i)\left[\frac{\partial}{\partial t_0} |(\exp(-iH(t - t_0)\alpha_i,\alpha_k)|^2\right]_{t_0=t} ,$$

is seen to represent the rate of change of the probability $(U_q(t)\alpha_k,\alpha_k)$ of finding $A = a_k$ (where a_k is the eigenvalue of α_k) immediately following a reduction over the α_k. When a system decays exponentially out of an excited state, in contrast, the initial rate of decay is nonzero. Since the above quantity is always zero for Hermitian H, exponential decay can never strictly occur in quantum mechanics.

The above result seems to me to rule out the possibility of obtaining satisfactory spontaneous postulates which preserve the quantum statistics, for according to such postulates any quantity which changes in time is unobservable, which does not appear to be the case. However, the requirement to "preserve the quantum statistics," which was intended to give precise expression to the notion of "empirical equivalence," is perhaps stronger than it need be. Certainly a spontaneous projection postulate might fail to preserve the quantum statistics and yet be indistinguishable from conventional quantum mechanics "in the accelerator," either because the relevant differences between $(U_q(t)\alpha_k,\alpha_k)$ and $(U_s(t)\alpha_k,\alpha_k)$ are too small, or appear for too short a time, or any one of a host of other possibilities. So it remains an open question whether there are satisfactory spontaneous projection postulates which fail to preserve the quantum statistics but which are "actually experimentally indistinguishable" from conventional mechanics. But we shall have to wait and examine each candidate as it appears, I suspect, for "actually experimentally indistinguishable" is not a notion that lends itself for the kind of armchair investigation that we have carried out here.

References

Cartwright, N. The only real probabilities in quantum mechanics. In P. D. Asquith & I. Hacking (Eds.), *PSA 1978* (Vol. 1). East Lansing, Mich.: Philosophy of Science Association, 1978.

Cartwright, N. *Classical families of observables: Towards an observer free quantum mechanics*. Paper presented at Boston University as part of the Boston Colloquium Series, January 1979.

Goldberger, M. L., & Watson, K. M. *Collision theory*. New York: Wiley, 1964.

Von Neumann, J. *Mathematical foundations of quantum mechanics*. Princeton, N.J.: Princeton University Press, 1955.

QUANTUM LOGIC CALISTHENICS

John E. Shively
Lawrence Berkeley Laboratory
University of California

The purpose of this paper[1] is to display an example of a nondistributive logic that avoids all the undesirable properties (given later in Table 2) of the usual very special case.[2] Though it is well known that modular lattices fail to distribute (cf. MacLane & Birkhoff, 1967, chap. 14, sec. 4, p. 490), it is useful to display a simple nontrivial example as in Table 3 (discussed below). These examples are given in terms of the complemented modular lattice of finite perpendicular[3] projection operators and their intersections, spans, and complements. Some properties of these operators are given in Table 1.

Let P and Q be two perpendicular projections onto the subspaces $\mathcal{H}_P$ and $\mathcal{H}_Q$ of a linear space $\mathcal{H}$.

The projection $X = P \wedge Q$ onto the intersection (cf. Halmos, 1967, problem 96, p. 58, and hint 96, p. 152) of $\mathcal{H}_P$ and $\mathcal{H}_Q$ is defined by

$$X = P \wedge Q \equiv \lim_{n \to \infty} (PQP)^n$$

and is thought to correspond to the logical connective 'and'.

The projection $S = P \vee Q$ onto the span (cf. Halmos, 1967, solution 96, p. 257) of $\mathcal{H}_P$ and $\mathcal{H}_Q$ is defined by a De Morgan's law

$$S = P \vee Q \equiv \overline{\overline{P} \quad \overline{Q}}$$

and is thought to correspond to the logical connective 'or'. The bar means orthogonal complement, $\overline{P} = I - P$, which is thought to correspond to the logical negation 'not'.

Studies in the Foundations of Quantum Mechanics, pp. 165-172

The projections are thought to correspond to propositions. These projections obey the equalities displayed in Table 1.

TABLE 1

Properties of Perpendicular Projection Operators

$$P \vee Q = Q \vee P \qquad P \wedge Q = Q \wedge P$$
$$PX = X \qquad QX = X$$
$$XP = X \qquad XQ = X$$
$$PS = P \qquad QS = Q$$
$$SP = P \qquad SQ = Q$$
$$\bar{P}X = 0 \qquad \bar{Q}X = 0$$
$$X\bar{P} = 0 \qquad X\bar{Q} = 0$$
$$\bar{P}S = S - P \qquad \bar{Q}S = S - Q$$
$$(S - P)^2 = S - P \qquad (S - Q)^2 = S - Q$$
$$S\bar{P} = S - P \qquad S\bar{Q} = S - Q$$

$$SX = X \qquad \bar{P} \wedge P = 0 \qquad \bar{Q} \wedge Q = 0$$
$$XS = X \qquad P \wedge \bar{P} = 0 \qquad Q \wedge \bar{Q} = 0$$
$$S \wedge X = X \qquad \bar{P} \vee P = I \qquad \bar{Q} \vee Q = I$$
$$S \vee X = S \qquad P \vee \bar{P} = I \qquad Q \vee \bar{Q} = I$$

$$P \vee Q + \bar{P} \wedge \bar{Q} = I \qquad (P - Q)^2 + (I - P - Q)^2 = I$$
$$P \wedge Q + \bar{P} \vee \bar{Q} = I \qquad (PQ - QP)^2 = (P - Q)^4 - (P - Q)^2$$
$$P \wedge P = P \qquad P \vee P = P$$
$$P \wedge (Q \wedge R) = (P \wedge Q) \wedge R \qquad P \vee (Q \vee R) = (P \vee Q) \vee R$$
$$P \wedge (P \vee Q) = P \qquad P \vee (P \wedge Q) = P$$

As an example of the method of computation, an intersection is calculated next.

Let

$$P = \begin{bmatrix} 1/2 & 0 & 0 & 1/2 \\ 0 & 0 & 0 & 0 \\ 0 & 0 & 1 & 0 \\ 1/2 & 0 & 0 & 1/2 \end{bmatrix} \quad \text{and} \quad Q = \begin{bmatrix} 1/3 & 1/3 & 0 & 1/3 \\ 1/3 & 1/3 & 0 & 1/3 \\ 0 & 0 & 1 & 0 \\ 1/3 & 1/3 & 0 & 1/3 \end{bmatrix} .$$

Then

$$PQP = \begin{bmatrix} 1/3 & 0 & 0 & 1/3 \\ 0 & 0 & 0 & 0 \\ 0 & 0 & 1 & 0 \\ 1/3 & 0 & 0 & 1/3 \end{bmatrix}$$

and

$$(PQP)^2 = \begin{bmatrix} 2/9 & 0 & 0 & 2/9 \\ 0 & 0 & 0 & 0 \\ 0 & 0 & 1 & 0 \\ 2/9 & 0 & 0 & 2/9 \end{bmatrix} .$$

Thus, by induction,

$$(PQP)^n = \begin{bmatrix} h & 0 & 0 & h \\ 0 & 0 & 0 & 0 \\ 0 & 0 & 1 & 0 \\ h & 0 & 0 & h \end{bmatrix} , \quad \text{where} \quad h = \frac{2^{n-1}}{3^n} .$$

Hence, by continuity and limits,

$$X = P \wedge Q = \lim_{n \to \infty} (PQP)^n = \begin{bmatrix} 0 & 0 & 0 & 0 \\ 0 & 0 & 0 & 0 \\ 0 & 0 & 1 & 0 \\ 0 & 0 & 0 & 0 \end{bmatrix} ,$$

since $\lim_{n \to \infty} \frac{2^{n-1}}{3^n} = 0.$

The usual very special case can be constructed from the projection

$$A(\theta) = \begin{bmatrix} \cos^2\theta & \sin\theta\cos\theta \\ \sin\theta\cos\theta & \sin^2\theta \end{bmatrix}$$

which is the perpendicular projection onto the line inclined at an angle θ from the first basis vector toward the second basis vector.

Note that

$$A^2(\theta) = A(\theta) \quad \text{and} \quad \mathrm{Tr}\, A(\theta) = 1 \ .$$

Define

$$P = A(\pi/4) = \begin{bmatrix} 1/2 & 1/2 \\ 1/2 & 1/2 \end{bmatrix}$$

$$Q = A(0) = \begin{bmatrix} 1 & 0 \\ 0 & 0 \end{bmatrix}$$

$$R = A(\pi/2) = \begin{bmatrix} 0 & 0 \\ 0 & 1 \end{bmatrix} .$$

This example not only illustrates the failure of the distributive law

$$P \wedge (Q \vee R) \neq (P \wedge Q) \vee (P \wedge R)$$

but also obeys all the undesirable special case properties listed in Table 2.

The simple nontrivial example displayed in Table 3 violates all the undesirable special case properties listed in Table 2 and demonstrates not only the failure (cf. Halmos, 1958, chap. 1, sec. 12, problem 5, p. 19) of the distributive law for the complemented (cf. MacLane & Birkhoff, 1967, chap. 14, sec. 8, p. 499) modular lattice of finite perpendicular projections

$$P \wedge (Q \vee R) \neq (P \wedge Q) \vee (P \wedge R) \ ,$$

but also the validity of the modified distributive law true for any modular lattice

$$P \wedge (Q \vee (P \wedge R)) = (P \wedge Q) \vee (P \wedge R) \ .$$

TABLE 2

Special Case Properties

$P \wedge Q = 0$	$Q \vee (P \wedge R) = Q$
$P \wedge R = 0$	$P \wedge (Q \vee R) = P$
$Q \wedge R = 0$	$(P \wedge Q) \vee (P \wedge R) = 0$
$Q \vee R = I$	$P \wedge (Q \vee (P \wedge R)) = 0$
$I - Q = R$	

A computational shortcut[4] is indicated by listing the ranges of the various projections in Table 4. Observe that the number of variables listed for each projection is equal to its trace. A little practice enables one to construct intersection and span projection matrices directly from the range variables, since the columns of a projection generate its range.

Perhaps a more fruitful approach to the lattice versus logic paradox would be to replace span with interpolation for the connective `or'.

Given two projections P and Q, one can construct (cf. Kato, 1966, problem 4.13, p. 34) a family of interpolated projections P(t) by

$$P(t) = \frac{1}{2}\left\{ I + (2P - I + 2t(Q - P)) \cdot (I - 4t(1 - t)R)^{-1/2} \right\}$$

where $R = (P - Q)^2$, provided spr $R < 1$.

Then $P(0) = P$ and $P(1) = Q$ and $P(t)$ is a projection in between P and Q for $0 < t < 1$.

TABLE 3

A Nontrivial Nondistributive Example of Finite Perpendicular Projection Operators

$$P = \begin{bmatrix} 1/2 & 0 & 0 & 1/2 \\ 0 & 1 & 0 & 0 \\ 0 & 0 & 1 & 0 \\ 1/2 & 0 & 0 & 1/2 \end{bmatrix}$$

$$Q = \begin{bmatrix} 1/2 & 0 & 1/2 & 0 \\ 0 & 1/2 & 0 & 1/2 \\ 1/2 & 0 & 1/2 & 0 \\ 0 & 1/2 & 0 & 1/2 \end{bmatrix}$$

$$R = \begin{bmatrix} 1/2 & 0 & 0 & -1/2 \\ 0 & 1/2 & -1/2 & 0 \\ 0 & -1/2 & 1/2 & 0 \\ -1/2 & 0 & 0 & 1/2 \end{bmatrix}$$

$$P \wedge Q = \begin{bmatrix} 1/4 & 1/4 & 1/4 & 1/4 \\ 1/4 & 1/4 & 1/4 & 1/4 \\ 1/4 & 1/4 & 1/4 & 1/4 \\ 1/4 & 1/4 & 1/4 & 1/4 \end{bmatrix}$$

$$P \wedge R = \begin{bmatrix} 0 & 0 & 0 & 0 \\ 0 & 1/2 & -1/2 & 0 \\ 0 & -1/2 & 1/2 & 0 \\ 0 & 0 & 0 & 0 \end{bmatrix}$$

$$Q \wedge R = \begin{bmatrix} 1/4 & -1/4 & 1/4 & -1/4 \\ -1/4 & 1/4 & -1/4 & 1/4 \\ 1/4 & -1/4 & 1/4 & -1/4 \\ -1/4 & 1/4 & -1/4 & 1/4 \end{bmatrix}$$

(Table 3--continued)

$$Q \vee R = \begin{bmatrix} 1 & 0 & 0 & 0 \\ 0 & 1 & 0 & 0 \\ 0 & 0 & 1/2 & -1/2 \\ 0 & 0 & -1/2 & 1/2 \end{bmatrix}$$

$$P \wedge (Q \vee R) = \begin{bmatrix} 1/3 & 0 & -1/3 & 1/3 \\ 0 & 1 & 0 & 0 \\ -1/3 & 0 & 1/3 & -1/3 \\ 1/3 & 0 & -1/3 & 1/3 \end{bmatrix}$$

$$(P \wedge Q) \vee (P \wedge R) = \begin{bmatrix} 1/4 & 1/4 & 1/4 & 1/4 \\ 1/4 & 3/4 & -1/4 & 1/4 \\ 1/4 & -1/4 & 3/4 & 1/4 \\ 1/4 & 1/4 & 1/4 & 1/4 \end{bmatrix}$$

$$Q \vee (P \wedge R) = \begin{bmatrix} 3/4 & 1/4 & 1/4 & -1/4 \\ 1/4 & 3/4 & -1/4 & 1/4 \\ 1/4 & -1/4 & 3/4 & 1/4 \\ -1/4 & 1/4 & 1/4 & 3/4 \end{bmatrix}$$

$$P \wedge (Q \vee (P \wedge R)) = \begin{bmatrix} 1/4 & 1/4 & 1/4 & 1/4 \\ 1/4 & 3/4 & -1/4 & 1/4 \\ 1/4 & -1/4 & 3/4 & 1/4 \\ 1/4 & 1/4 & 1/4 & 1/4 \end{bmatrix}$$

$$I = \begin{bmatrix} 1 & 0 & 0 & 0 \\ 0 & 1 & 0 & 0 \\ 0 & 0 & 1 & 0 \\ 0 & 0 & 0 & 1 \end{bmatrix}$$

TABLE 4

Ranges of the Projection Operators

P	Q	R	$P \wedge Q$	$P \wedge R$	$Q \wedge R$	$Q \vee R$	$P \wedge (Q \vee R)$
α	α	α	α	0	α	α	α
β	β	β	α	γ	$-\alpha$	β	β
γ	α	$-\beta$	α	$-\gamma$	α	γ	$-\alpha$
α	β	$-\alpha$	α	0	$-\alpha$	$-\gamma$	α

$(P \wedge Q) \vee (P \wedge R)$	$Q \vee (P \wedge R)$	$P \wedge (Q \vee (P \wedge R))$	I
α	α	α	α
$\alpha + \gamma$	$\beta + \gamma$	$\alpha + \gamma$	β
$\alpha - \gamma$	$\alpha - \gamma$	$\alpha - \gamma$	γ
α	β	α	δ

Notes

[1]This paper is based on the author's quantum logic seminar at Stanford University, November 13, 1978.

[2]Quantum logic seminar delivered by J. Beehner at Stanford University, November 7, 1977.

[3]Null space orthogonal to range; hence, if real, then symmetric.

[4]Suggested by Professor Leonard Monk, November 13, 1978.

References

Halmos, P. R. *Finite-dimensional vector spaces* (2nd ed.). Princeton, N.J.: Van Nostrand, 1958.

Halmos, P. R. *A Hilbert space problem book*. Princeton, N.J.: Van Nostrand, 1967.

Kato, T. *Perturbation theory for linear operators*. New York: Springer-Verlag, 1966.

MacLane, S., & Birkhoff, G. *Algebra*. New York: Macmillan, 1967.

A NEW PROOF OF THE IMPOSSIBILITY OF HIDDEN VARIABLES USING THE PRINCIPLES OF EXCHANGEABILITY AND IDENTITY OF CONDITIONAL DISTRIBUTIONS

Patrick Suppes and Mario Zanotti
Stanford University

The main purpose of this paper is to provide a new proof of the impossibility of local theories of hidden variables based on simpler and more general assumptions than those used by Bell (1964, 1966), Wigner (1970), and Suppes and Zanotti (1976). In particular, no assumptions requiring specific quantum mechanical calculations are required. They are replaced by the principle of exchangeability and the principle of identical conditional distributions given the hidden variable.

The results may be most easily discussed in terms of a system of two spin-1/2 particles initially in the singlet state, but generalizations to other quantum mechanical systems of a similar nature are apparent. In essential terms, the analysis of Bell, stated in a particularly clear form by Wigner, depends upon the following assumptions, which we state in intuitive form. (Mathematically explicit axioms are formulated in Appendix A.)

1. Axial symmetry. For any direction of the measuring apparatus the expected spin is 0, where spin is measured by +1 and -1 for spin +1/2 and spin -1/2, respectively. Further, the expected product of the spin measurements is the same for different orientations of the measuring apparatuses, as long as the angle between the measuring apparatuses remains the same.

2. Opposite measurement for same orientation. The correlation between the spin measurements is -1 if the two measuring apparatuses have the same orientation.

3. Independence of λ. The expectation of any function of λ is independent of the orientation of the measuring apparatus.

Studies in the Foundations of Quantum Mechanics, pp. 173-191

4. Locality. The spin measurement obtained with one apparatus is independent of the orientation of the other measuring apparatus.

5. Determinism. Given λ and the orientation of the measuring apparatus, the result of the spin measurement is determined uniquely. In other words, the conditional variance of the spin measurement, given λ and the direction of the measuring apparatus, is 0.

Suppes and Zanotti (1976) weaken the last assumption of determinism to conditional statistical independence, that is, to the assumption that the expectation of the product of the spin measurements, given λ and the orientation of the measuring apparatuses, is equal to the product of the expectations under the same conditions. The heart of their argument is to show that statistical independence given λ and a correlation of -1 between the measurements implies determinism.

In the present paper we simplify these arguments drastically by using directly de Finetti's principle of exchangeability and the assumption of a negative correlation to prove the nonexistence of a hidden variable theory of the kind that is usually called objective and local. What is significant about our argument, in our judgment, is that it eliminates almost all of the quantum mechanical details from the argument for the nonexistence of hidden variable theories. The conditions that have to be satisfied are quite general. The argument shows how weak the conditions may be that lead to nonexistence of objective local hidden variables.

We need to say something about the principle of exchangeability because it is a principle of symmetry that has not been used in physics, to our knowledge, but has played an important role in subjective theories of probability. The principle was introduced by de Finetti to provide a natural alternative in the subjective theory of probability to objective theories of independence. (We emphasize that the conflicts between objective and subjective theories of probability do not play a role in our own argument here. The principle of exchangeability enters in a formal way and does not require any subjective interpretation, although we are in favor of this broader viewpoint.)

Here is a simple example to illustrate the principle. Suppose we flip 10 times a coin that is known to be biased, and we know artificially that the bias is either .75 in favor of heads or .25 in favor of heads. Then the flips will not be independent because the outcomes of preceding flips will provide information about the probability of a head on the next flip. On the other hand, given the number of heads that occur in 10 flips, the trials on which the heads occur is of no importance. In other words, we have permutational invariance in the sense that the probability of a sequence of 10 outcomes with a fixed number of heads is the same regardless of exactly which trials heads occurred on. Thus, for example, if three heads occurred in 10 flips, then it does not matter whether these three heads occurred on the first three trials, or on the first, seventh, and tenth, etc. Notice how much the principle of exchangeability reduces the number of probabilities that have to be determined. In the present case, instead of considering 2^{10} sequences of possible outcomes, we can reduce this to just 11, the probabilities of 0 to 10 heads, and because of conservation of probability we in fact need to fix only the probability of 10 with the 11th then determined, which is a very considerable reduction of parameters from 2^{10}. The kind of symmetry that is behind the principle of exchangeability and how it applies to the spin experiments should be clear from this example. Exchangeability is at the heart of the symmetry implied by the physics of the spin experiments.

The requirement of exchangeability is uncontroversial. This is not so for the associated requirement of identity of conditional distribution for the two measurements when the apparatuses have the same orientation. No trace of this axiom is to be found in the papers of Bell and Wigner cited above, or in the recent excellent survey paper by Clauser and Shimony (1978), which covers most of the earlier literature. But identity of conditional distribution seems to be a reasonable symmetry requirement, because the hidden variable λ should have the same _expected_ causal effect on both particles. Obviously, it is consistent with this axiom that the _actual_ causal effects of λ on the particles are different, which is certainly true in the paradigm stochastic experiments on coin flipping.

Especially in relation to the earlier literature on hidden variables, there are two points to be made. As a principle of symmetry, the existence of identical conditional distributions is a different and stronger principle than that of exchangeability--equivalence of the two is true only for infinite sequences of random variables. Second, the principle of identity of conditional distributions is in itself a wholly reasonable principle of symmetry. To take a very elementary example, consider throwing out a pair of fair dice. They have the same conditional distributions on each throw, for there is no principle of difference to make the distributions different, but, of course, we expect the actual outcomes to be different most of the time--five-sixths of the time to be exact. And so it is with spin-1/2 particles initially in the singlet state, and a number of other particle systems that are described by Clauser and Shimony as suitable for testing hypotheses about hidden variables.

The fundamental point is that it is the strong symmetry of the principle of identity of conditional distributions that is the main tool for proving the nonexistence of hidden variables.

I. A GENERAL PROBABILISTIC THEOREM ON HIDDEN VARIABLES

In our earlier paper on hidden variable theories (Suppes & Zanotti, 1976), we proved, in a general probabilistic lemma, (i) negative correlation of -1 between two random variables X and Y and (ii) their conditional independence given a random variable λ, that is,

$$E(XY|\lambda) = E(X|\lambda)E(Y|\lambda) \ ,$$

imply that λ completely determines the values of X and Y.

In the present context, we establish a necessary and sufficient condition for two-valued exchangeable random variables to have 'identical' causes in the sense of conditional expectation.

THEOREM 1. *Let* X *and* Y *be two-valued random variables, for definiteness, with possible values* 1 *and* -1, *and with positive variances, i.e.,* $\sigma(X),\sigma(Y) > 0$. *In addition, let* X *and* Y *be exchangeable, i.e.,*

$$P(X = 1, Y = -1) = P(X = -1, Y = 1) \ .$$

Then a necessary and sufficient condition that there exist a hidden variable $\underline{\lambda}$ *such that* $E(XY|\underline{\lambda} = \lambda) = E(X|\underline{\lambda} = \lambda)E(Y|\underline{\lambda} = \lambda)$ *and* $E(X|\underline{\lambda} = \lambda) = E(Y|\underline{\lambda} = \lambda)$ *for every value* λ *(except possibly on a set of measure zero) is that the correlation of* X *and* Y *be nonnegative*.

Proof. We first prove an elementary lemma giving a necessary and sufficient condition for the correlation, $\rho(X,Y)$, of X and Y to be negative.

LEMMA 1. *Under the hypothesis of Theorem* 1,

$$\rho(X,Y) < 0 \text{ iff } P(X = 1, Y = -1) > P(X = 1)P(Y = -1) .$$

Proof. Since $\sigma(X) = \sigma(Y) > 0$, we have $\rho(X,Y) < 0$ if and only if $\text{cov}(X,Y) < 0$. Next we compute the covariance.

$$\begin{aligned} \text{cov}(X,Y) &= E(XY) - E(X)E(Y) = E(XY) - E^2(X) \\ &= P(X = 1, Y = 1) + P(X = -1, Y = -1) \\ &\quad - 2P(X = 1, Y = -1) - P^2(X = 1) \\ &\quad - P^2(X = -1) + 2P(X = 1)P(X = -1) \\ &= 4P(X = 1)P(X = -1) - 4P(X = 1, Y = -1) . \end{aligned}$$

Thus, $\text{cov}(X,Y) < 0$ if and only if $P(X = 1, Y = -1) > P(X = 1)P(X = -1)$, as desired.

We next prove half of our theorem, namely, that under the hypotheses stated, conditional independence with identical conditional distributions given a hidden variable $\underline{\lambda}$ implies nonnegative correlation.

First, we note, under our assumptions,

$$\begin{aligned} P(X = 1) = P(Y = 1) &= \int P(X = 1|\underline{\lambda} = \lambda)\, dP(\lambda) \\ &= \int P(Y = 1|\underline{\lambda} = \lambda)\, dP(\lambda) \\ &= \pi , \end{aligned}$$

where, here and in what follows, the integration is from $-\infty$ to ∞. (Actually, it can be shown that for random variables X and Y with only two values, it suffices to suppose λ has only a fixed finite set of values--for either the proof of sufficiency or of necessity.)

Next, we examine

$$P(X = 1, Y = 1) = \int P(X = 1, Y = 1 | \underline{\lambda} = \lambda)\, dP(\lambda)$$
$$= \int P(X = 1 | \underline{\lambda} = \lambda) P(Y = 1 | \underline{\lambda} = \lambda)\, dP(\lambda)$$
$$= \int P^2(X = 1 | \underline{\lambda} = \lambda) dP(\lambda) ,$$

under our hypothesis of indentical conditional distributions for X and Y. Now we have from the above by obvious algebraic manipulation

$$P(X = 1, Y = 1) = \int (\pi - P(X = 1 | \underline{\lambda} = \lambda))^2 dP(\lambda)$$
$$- \int \pi^2 dP(\lambda) + 2 \int \pi P(X = 1 | \underline{\lambda} = \lambda)\, dP(\lambda)$$
$$= \int (\pi - P(X = 1 | \underline{\lambda} = \lambda)^2\, dP(\lambda) + \pi^2$$
$$= \int (\pi - P(X = 1 | \underline{\lambda} = \lambda)^2\, dP(\lambda)$$
$$+ P(X = 1) P(Y = 1) ,$$

and since the first term is clearly nonnegative, we have

$$P(X = 1, Y = 1) \geq P(X = 1) P(Y = 1) ,$$

and by exactly similar argument

$$P(X = -1, Y = -1) \geq P(X = -1) P(Y = -1) ,$$

whence, adding inequalities and using conservation of probability, we have

$$1 - 2P(X = 1, Y = -1) \geq 1 - 2P(X = 1) P(Y = -1) ,$$

and so

$$P(X = 1, Y = -1) \leq P(X = 1) P(Y = -1) .$$

Consequently, by Lemma 1,

$$\rho(X,Y) \geq 0 ,$$

as desired.

We now prove that a nonnegative correlation of X and Y implies existence of a hidden variable that will make X and Y be conditionally independent and have identical conditional distributions. We give an elementary proof that such a hidden variable exists. A more general proof using Choquet's theorem is given in Appendix B.

Let $\alpha_1 = P(X = 1)$, $\alpha_2 = P(X_1 = 1, Y = 1)$. Then α_1 and α_2 uniquely determine the distribution for the pair of exchangeable random variables X and Y and satisfy conditions:

(i) $\alpha_2 \geq 0$;

(ii) $\alpha_1 - \alpha_2 \geq 0$;

(iii) $1 - 2\alpha_1 + \alpha_2 \geq 0$;

Moreover, $\rho(X,Y) \geq 0$ if and only if the following condition is satisfied:

(iv) $\alpha_2 - \alpha_1^2 \geq 0$.

Note that condition (iii) is implied by condition (iv).

We now introduce a two-valued random variable $\underline{\lambda}$ and show that the conditional probabilities $P(X = 1|\underline{\lambda} = \lambda_1) = P(Y = 1|\underline{\lambda} = \lambda_1)$, $P(X = 1|\underline{\lambda} = \lambda_2) = P(Y = 1|\underline{\lambda} = \lambda_2)$, and $P(\underline{\lambda} = \lambda_1)$ may be assigned values compatible with α_1 and α_2.

By hypothesis of conditional independence and identity of conditional distributions for X and Y:

$$\begin{aligned} P(X = 1, Y = 1) &= P(X = 1, Y = 1|\underline{\lambda} = \lambda_1)P(\underline{\lambda} = \lambda_1) \\ &\quad + P(X = 1, Y = 1|\underline{\lambda} = \lambda_2)P(\underline{\lambda} = \lambda_2) \\ &= P^2(X = 1|\underline{\lambda} = \lambda_1)P(\underline{\lambda} = \lambda_1) \\ &\quad + P^2(X = 1|\underline{\lambda} = \lambda_2)P(\underline{\lambda} = \lambda_2) . \end{aligned}$$

We now set

$$P(X = 1 | \underline{\lambda} = \lambda_1) = \frac{1 - \sqrt{1 - 4(\alpha_1 - \alpha_2)}}{2}$$

$$P(X = 1 | \underline{\lambda} = \lambda_2) = \frac{1 + \sqrt{1 - 4(\alpha_1 - \alpha_2)}}{2}$$

$$P(\underline{\lambda} = \lambda_1) = \frac{1}{\sqrt{1 - 4(\alpha_1 - \alpha_2)}} \left(\frac{1 + \sqrt{1 - 4(\alpha_1 - \alpha_2)}}{2} - \alpha_1 \right) .$$

Then

$$\alpha_1 = P(X = 1) = P(X = 1 | \underline{\lambda} = \lambda_1) P(\underline{\lambda} = \lambda_1) + P(X = 1 | \underline{\lambda} = \lambda_2) P(\underline{\lambda} = \lambda_2)$$

and

$$\alpha_2 = P(X = 1, Y = 1) = P^2(X = 1 | \underline{\lambda} = \lambda_1) P(\underline{\lambda} = \lambda_1) + P^2(X = 1 | \underline{\lambda} = \lambda_2) P(\underline{\lambda} = \lambda_2) ,$$

as is easily checked. Moreover, it is also easy to check that

$$0 \leq P(X = 1 | \underline{\lambda} = \lambda_1), P(X = 1 | \underline{\lambda} = \lambda_2), P(\underline{\lambda} = \lambda_1) \leq 1 ,$$

and the proof of Theorem 1 is complete.

II. APPLICATION TO QUANTUM MECHANICS

Because the application of hidden variable arguments to a system of two spin-1/2 particles initially in the singlet state is now so familiar, and because we have treated it very formally from a probabilistic standpoint in our earlier article (Suppes & Zanotti, 1976), we shall keep the discussion informal here, but it is obvious how it can be made as formal as desired, and this is done in Appendix A.

As we argued in our earlier paper, the notion of locality of an objective hidden variable theory is essentially a principle of conditional statistical independence. Let us now introduce the term _strict_ _local_ hidden variable theory for a hidden variable theory that satisfies the following two principles: (i) the principle of conditional statistical independence and (ii) the principle of identity of conditional distributions for particles that satisfy the principle of exchangeability.

Before stating the theorem about strict local hidden variable theories, we want to take note of a point that Clauser and Shimony (1978) make. Although in the Bell arguments it is assumed that when the measuring apparatus is oriented the same in the spin experiments the correlation observed will be -1, in fact, because of the absence of completely efficient measuring apparatus, the actual observed correlation will be definitely negative but not -1. The theorem takes account of this consideration.

THEOREM 2. _Given (i) the experimentally determined negative correlation of the spin of two spin-1/2 particles that were initially in the singlet state and (ii) satisfaction of the principle of exchangeability by the probabilities of spin measurements for the two particles, there can be no strict local hidden variable theory of the phenomena._

The proof of this theorem follows at once from the hypotheses and Theorem 1.

This negative result raises the question of what kind of theories we are content to have for symmetric phenomena with correlations. What the theorem shows is that no fully satisfactory causal theory is possible for the standard quantum mechanical experiments now discussed extensively in connection with local hidden variable theories.

Let us review the four principles we have been discussing that cannot be simultaneously satisfied.

(i) Principle of exchangeability: a phenomenological principle of symmetry.

(ii) Negative correlations between random variables satisfying the principle of exchangeability: a phenomenological result.

(iii) Identity of conditional distributions given a hidden variable for particles satisfying the principle of exchangeability: a theoretical principle of symmetry.

(iv) Conditional statistical independence given a hidden variable for particles satisfying the principle of exchangeability: a theoretical principle of locality of causes.

What Theorem 2 shows is that several different kinds of quantum mechanical phenomena cannot simultaneously satisfy all four of these principles, and consequently without any detailed quantum mechanical calculations at all there can be no strict local hidden variable theory.

The question then is what attitude to take about the possibility of theories that are not strict. Much of the physical discussion has argued for the principle of locality as reflected in (iv), but in our view this is _the_ principle of the four to be given up, and in our earlier paper we discussed simple examples of stochastic processes that did not satisfy this principle and yet had an apparently satisfactory causal analysis, at least a satisfactory causal analysis up to a certain point--obviously not up to a point that satisfies the principles of conditional independence and of identity of conditional distributions for particles that obey exchangeability.

We modify that simple example of coin flipping so as not to have a perfect negative correlation of -1. Let X be the random variable whose values are heads (+1) or tails (-1) at detector 1 and let Y be defined similarly at detector 2. Then we set

$$P(X = 1, Y = -1 \mid \underline{\lambda} = \lambda) = P(X = -1, Y = 1 \mid \underline{\lambda} = \lambda) = \lambda > \frac{1}{4}$$

$$P(X = 1, Y = 1 \mid \underline{\lambda} = \lambda) = P(X = -1, Y = -1 \mid \underline{\lambda} = \lambda) = \frac{1}{2} - \lambda .$$

Note that $1/4 < \lambda < 1/2$ by hypothesis, and so $1/4 < E(\underline{\lambda}) < 1/2$. Thus,

$$\begin{aligned} P(X = 1, Y = -1) = P(X = -1, Y = 1) &= \int \lambda \, dP(\lambda) \\ &= E(\lambda) \end{aligned}$$

$$\begin{aligned} P(X = 1, Y = 1) &= P(X = -1, Y = -1) \\ &= \int \left(\frac{1}{2} - \lambda \right) dP(\lambda) \\ &= \frac{1}{2} - E(\lambda) \end{aligned}$$

$$\begin{aligned} P(X = 1) &= P(Y = 1) \\ &= \int \frac{1}{2}\, dP(\lambda) \\ &= \frac{1}{2} . \end{aligned}$$

Consequently,

$$E(X) = E(Y) = 0$$

$$\sigma^2(X) = \sigma^2(Y) = 1 ,$$

and thus

$$\rho(X,Y) = E(XY) = 1 - 4E(\lambda) < 0 .$$

The causal setup is supposed to operate as follows. A device is at work at the source representing the physical embodiment of the hidden variable $\underline{\lambda}$. When λ is the outcome, then the orientation of the coins is determined by a random device using the probability λ according to the equation given above. Use of this random device blocks any further causal anaysis--on our hypothesis of its genuine randomness.

Yet in this example we feel partly satisfied about the causal analysis because the orientation of each coin is fixed by a random device at the source. But we can still feel unsatisfied because of the phenomenological negative correlation. From a causal standpoint we can feel that something is missing that a more complete analysis would permit us to introduce, for example, additional phenomenological variables as well as additional theoretical variables, so that particles put in `identical circumstances′ will, in terms of their probabilistic behavior if not in terms of their behavior on each occasion, not only behave the same but all of the information required to predict their behavior will be embodied in the causal analysis made and, thus, conditional statistical independence will hold. But for such independence to hold we must be able to introduce further variables to eliminate the negative correlations and contravene our hypothesis of genuine randomness.

When we cannot make such improvements, we must face the situation, which we do not always seem willing to do--that for such phenomena we do not have a fully satisfactory causal theory. It seems to us that this is the situation in quantum mechanics. Our general ideas of strict probabilistic causality cannot be satisfied, not to speak of deterministic causation, by quantum mechanics.

Another way of putting it is that when we sacrifice the principle of conditional statistical independence we are saying that we are satisfied with a weaker, that is, a nonstrict, causal theory. No hidden variable theory not satisfying this principle can hope to give us all of the information we would expect to have about the phenomena in question. What is important about the quantum mechanical case is that it seems to be a genuine example of being able to have such a restricted stochastic hidden variable theory. We have termed the principle of conditional statistical independence a principle of locality, but it should be evident that it can be violated by phenomena as in the case of the simple coin-flipping example discussed and yet no hidden action at a distance takes place. The proper interpretation of the locality is that the phenomena cannot be deterministically localized if this principle is not satisfied. It is, of course, an extrapolation to talk about determinism here. As we proved in our previous article, however, in the case of perfect negative correlations, conditional statistical independence does indeed imply determinism, and this, it seems to us, is the reason for baptizing it as a principle of locality.

What remains is the possibiliy of weak stochastic hidden variable theories satisfying principles (i)-(iii). We have proved that there cannot be strict local hidden variable theories satisfying principles (i)-(iv) for quantum mechanics.

APPENDIX A

Formalization of Theorem 2

We formulated Theorem 2 only informally in the text. Here, in the spirit of our earlier paper, we make the mathematical statement explicit. We change, for the present purpose, the axioms stated in our earlier paper; for example, we now assume only negative correlation, not correlation -1.

The random variables we need are these, stated in terms of the concepts introduced at the beginning of the paper:

ω_I - the direction of orientation of measuring apparatus I;

ω_{II} - the direction of orientation of measuring apparatus II;

M_I - the spin measurement of apparatus I;

M_{II} - the spin measurement of apparatus II;

$\underline{\lambda}$ - the hidden variable.

The values of random variables ω_I and ω_{II} are direction vectors, for example, three-dimensional vectors normed to one, and the cosine of the angle between them is the scalar product $\omega_I \cdot \omega_{II}$, which is itself a new random variable. The values of random variables M_I and M_{II} are +1 and -1, for spin +1/2 and spin -1/2, respectively. Finally, we shall assume for simplicity of notation that $\underline{\lambda}$ is a real-valued random variable; but it could be vector-valued and not affect any of the theory, for the essential assumptions about the hidden variable $\underline{\lambda}$ are minimal.

DEFINITION 1. <u>A structure</u> $\langle \omega_I, \omega_{II}, M_I, M_{II}, \underline{\lambda} \rangle$ <u>of random variables is a</u> strict local hidden variable spin-1/2 system <u>if and only if the following axioms are satisfied:</u>

AXIOM 1. [<u>Exchangeability</u>]

$$E(M_I = 1, M_{II} = -1 \mid \omega_I = w_1, \omega_{II} = w_2)$$
$$= E(M_I = -1, M_{II} = 1 \mid \omega_I = w_1, \omega_{II} = w_2) .$$

AXIOM 2. [<u>Axial Symmetry</u>]

(i) $E(M_I|\omega_I) = E(M_{II}|\omega_{II}) = 0$;

(ii) <u>If we define for any direction vectors</u> w_I <u>and</u> w_{II} <u>and any real number</u> α

$$H(w_I,w_{II},\alpha) = [\omega_I = w_I,\omega_{II} = w_{II},w_I\cdot w_{II} = \alpha] ,$$

<u>then</u>

$$E(M_IM_{II}|H(w_I,w_{II},\alpha)) = E(M_IM_{II}|H(w_I',w_{II}',\alpha'))$$

<u>if</u> $\alpha = \alpha'$;

AXIOM 3. [<u>Negative Correlation for Same Orientation</u>] <u>If</u> $\alpha = 1$, <u>then</u>

$$\rho(M_IM_{II}|H(w_I,w_{II},\alpha)) < 0 ,$$

$$\sigma^2(M_I|H(w_I,w_{II},\alpha)) > 0 ,$$

$$\sigma^2(M_{II}|H(w_I,w_{II},\alpha)) > 0 ,$$

<u>where</u> ρ <u>is the conditional correlation and</u> σ^2 <u>is the conditional variance</u>.

AXIOM 4. [<u>Independence of</u> $\underline{\lambda}$] <u>For all functions</u> g <u>for which the expectations</u> $E(g(\lambda))$ <u>and</u> $E(g(\lambda)|\omega_I,\omega_{II})$ <u>are finite</u>,

$$E(g(\underline{\lambda})) = E(g(\underline{\lambda})|\omega_I,\omega_{II}) .$$

AXIOM 5. [<u>Locality: Independence of Orientation of Other Measuring Apparatus</u>]

$$E(M_I|\omega_I,\omega_{II},\underline{\lambda}) = E(M_I|\omega_I,\underline{\lambda}) ,$$

$$E(M_{II}|\omega_I,\omega_{II},\underline{\lambda}) = E(M_{II}|\omega_{II},\underline{\lambda}) .$$

AXIOM 6. [<u>Conditional Statistical Independence</u>]

$$E(M_IM_{II}|\omega_I,\omega_{II},\underline{\lambda}) = E(M_I|\omega_I,\omega_{II},\underline{\lambda})E(M_{II}|\omega_I,\omega_{II},\underline{\lambda}) .$$

AXIOM 7. [<u>Identity of Conditional Distributions</u>]

$$E(M_I = 1|\omega_I,\omega_{II},\underline{\lambda}) = E(M_{II} = 1|\omega_I,\omega_{II},\underline{\lambda})$$

Using the concepts introduced in Definition 1, we may formulate Theorem 2 more formally as follows.

THEOREM 2. [<u>Formal Version</u>] <u>There are no strict local hidden variable spin</u>-1/2 <u>systems in the sense of Definition</u> 1.

<u>Proof</u>. In view of Theorem 1, the axioms of Definition 1 are inconsistent, and thus there are no models of the axioms. The proof of inconsistency uses especially--as is obvious from the hypothesis of Theorem 1--Axioms 1, 3, 6, and 7.

APPENDIX B

General Proof of Existence of Hidden Variables

We give in this Appendix a general proof of the existence of a hidden variable when X and Y have nonnegative correlation and are exchangeable, with no special assumptions about the number of values of λ, as in the elementary proof given in the text for the second half of Theorem 1.

Let M denote the class of all probability measures induced by two exchangeable random variables X and Y on the Cartesian product $\{1,-1\} \times \{1,-1\}$ satisfying the conditions

$$P(X = 1, Y = 1) \geq P^2(X = 1), \text{ for all } P \in M .$$

Obviously, the set M is convex and compact. We now define: P *is an extreme point of* M *iff if* $P_1, P_2 \in M$, $0 < \alpha < 1$, $P_1 \neq P_2$, *and* $P = \alpha P_1 + (1 - \alpha)P_2$, *then* $\alpha = 0$ *or* $\alpha = 1$.

LEMMA 2. *Let* P *be an element of* M *such that* $P(X = 1, Y = 1) = P^2(X = 1)$. *Then* P *is an extreme point of* M.

Proof. Suppose not; that is, assume $P = \alpha P_1 + (1 - \alpha)P_2$ with $P_1 \neq P_2 \in M$ and $0 < \alpha < 1$. Then

$$\begin{aligned} P(X = 1, Y = 1) &= \alpha P_1(X = 1, Y = 1) \\ &\quad + (1 - \alpha)P_2(X = 1, Y = 1) \\ &\geq \alpha P_1^2(X = 1) + (1 - \alpha)P_2^2(X = 1) . \end{aligned}$$

Consider now the random variable

$$Z = HP_1(X = 1) + \bar{H}P_2(X = 1)$$

with $P(H) = \alpha$. Apply Jensen's inequality to $f(Z) = Z^2$ and we have

$$E^2(Z) \leq E(Z^2) ,$$

that is,

$$(\alpha P_1(X = 1) + (1 - \alpha)P_2(X = 1))^2$$
$$< \alpha P_1^2(X = 1) + (1 - \alpha)P_2^2(X = 1) ,$$

since equality is attained if and only if $P_1(X = 1) = P_2(X = 1)$. Combining, we obtain

$$P(X = 1, Y = 1) > (\alpha P_1(X = 1) + (1 - \alpha)P_2(X = 1))^2$$
$$= P^2(X = 1) .$$

Now let $\pi_{\{1,-1\}}$ be the set of all probability measures on $\{1,-1\}$. It is clear that for every $p \in \pi_{\{1,-1\}}$ the product measure $p \times p$ is an extreme point of M. The converse is also true. Namely, we have

$$\text{Extreme } (M) = \{p \times p: p \in \pi_{\{1,-1\}}\} .$$

We now show that each extreme point of M is a product measure. Suppose not; that is, assume there exists P, an extreme point of M, such that $P(X = 1, Y = 1) \neq P^2(X = 1)$. Clearly, $0 < P(X = 1) < 1$. Define

$$P_1(X = 1) = \frac{P(X = 1, Y = 1)}{P(X = 1)}$$

and

$$P_2(X = 1) = \frac{P(X = -1, Y = 1)}{P(X = -1)} .$$

Then

$$P(X = 1) = P(X = 1, Y = 1) + P(X = -1, Y = 1)$$
$$= P_1(X = 1)P(Y = 1) + P_2(X = 1)P(X = -1)$$
$$= \alpha P_1(X = 1) + (1 - \alpha)P_2(X = 1) ,$$

and so P is not an extreme point of M, contrary to our supposition. Applying Choquet's theorem (Meyer, 1966, p. 229), we have:

PROPOSITION: Each exchangeable probability measure P on the Cartesian product $\{1,-1\} \times \{1,-1\}$ satisfying

$$P(X = 1, Y = 1) \geq P(X = 1)^2$$

is of the form

$$P = \int_{[0,1]} p \times p \; dQ(p)$$

where Q is the distribution induced on the interval [0,1] by a hidden variable $\underline{\lambda}$.

References

Bell, J. S. On the Einstein-Podolsky-Rosen paradox. *Physics*, 1964, *1*, 195-200.

Bell, J. S. On the problem of hidden variables in quantum mechanics. *Reviews of Modern Physics*, 1966, *38*, 447-452.

Clauser, J. F., & Shimony, A. Bell's theorem: Experimental tests and implications. *Reports on Progress in Physics*, 1978, *41*, 1881-1927.

Meyer, P. A. *Probability and potential*. Waltham, Mass.: Blaisdell, 1966.

Suppes, P., & Zanotti, M. On the determinism of hidden variable theories with strict correlation and conditional statistical independence of observables. In P. Suppes (Ed.), *Logic and probability in quantum mechanics*. Dordrecht, The Netherlands: Reidel, 1976.

Wigner, E. P. On hidden variables and quantum mechanical probabilities. *American Journal of Physics*, 1970, *38*, 1005-1009.

CLASSICAL AND QUANTUM DECISION THEORY

Zoltan Domotor
University of Pennsylvania

I. BACKGROUND AND OBJECTIVES

One of the most challenging foundational topics of quantum mechanics is no doubt probability and measurement. While in classical physics one is to a great extent separable from the other and the possible connections are seen via error theory and ergodic theory, in quantum mechanics the two form a profound unity. Wild suggestions have been made in the past, for example, that quantum probability is actually classical or that it in fact does not exist; amateurism and controversy will probably continue longer than one cares to guess. The interpretation of quantum measurement is even more insidious.

One question to ask is, Why does one need probabilities at all, when, for example, quantum dynamics is actually quite innocuous and lives in great harmony with classical mechanics? The general answer is that one does not really need probabilities unless one wants to make decisions (concerning hypothesis testing and estimation). Probability is not an end product in itself, and one should not waste too much time trying to find out what it is. What we need to know is how it functions in particular modes of decision. The 'what question' regarding probability is still much alive even in classical contexts, as if there were a unique answer to it. It is known that probabilities come from a number of very different sources, commonly presented by (subjective) order relations, computational complexity functions, geometric inner products, and the like, and there is no hope for a unified starting point.

The aim of this paper is twofold. First, I want to suggest a fresh approach to decision theory, using two simple primitives: spaces and transitions between spaces.

Studies in the Foundations of Quantum Mechanics, pp. 193-222

Axioms concerning composition, product, conditionalization, and preference of transitions are flexible enough to accommodate the Bayesian dogma in its full generality and rigor in a wide variety of contexts. Second, I will show how classical decision theory applies without major changes to quantum measurement.

Transitions are profound and powerful functions with two aspects: measurement and transformation. In Part II, I show how probabilities and measurable maps fall out as derived notions. Part III displays the details of quantum decision and points out several analogies with classical reasoning. Orthocomplemented lattices and operator-valued probabilities fit nicely with the structure of decision. In Part IV, I deal with measurement of noncommuting observables. My main point (based on Naimark's theorem) is that, although Heisenberg's principle of uncertainty forbids any conjoint measurement of two noncommuting operators with limit-free accuracy, there is always a modified measurement (on an enlarged system) of another pair of commuting operators which gives optimal information about the noncommuting pair, with errors smaller than the ones dictated by Heisenberg. The results are essential to applications in optical communication, where quantum noise dominates other types of noise.

The theory does not depend on one's foundational idiosyncrasies concerning probability; it rests on the essence of decision.

This paper is a report on work in progress and is a revised version of the lectures given in Patrick Suppes' seminar on the foundations of quantum mechanics at Stanford during November 1975. The author is indebted to P. Suppes, J. Sneed, and M. Zanotti for many kinds of information, help, and stimulation.

II. STATISTICAL DECISION THEORY

The crucial structure of classical statistical decision theory is provided by the <u>composition</u>, <u>product</u>, <u>localization</u>, and <u>partial ordering</u> of <u>transitions</u>

$$X \xrightarrow{T} Y .$$

We think of X and Y as <u>measurable spaces</u>, allocated to particular, implicit (empirical or other) systems and impressed with observable (Boolean, Borel, or other) algebras of events which, when needed, are made explicitly visible by labels $\underline{B}_X$ and $\underline{B}_Y$. A transition T is a real-valued function

$$T : X \times \underline{B}_Y \longrightarrow \mathbb{R}$$

such that

(a) its first section $T(x) =_D T(x,\bullet) : \underline{B}_Y \longrightarrow \mathbb{R}$ is a probability measure on Y for all x in the sample space X;

(b) its second section $T(\bullet,B) : X \longrightarrow \mathbb{R}$ is a random variable on X for all B in $\underline{B}_Y$.

The algebraic approach and choice of notation resulted from much searching for the simplest means of presenting the theory. The abstract arrow for T suggests a probabilistic information transfer from the domain space X to the codomain space Y. Philosophically it indicates that to each possible state x one assigns a particular distribution of beliefs over the observational propositions regarding x.

The most fundamental operation on transitions is (sequential) composition. It is introduced by the formal rule

$$\frac{X \xrightarrow{T} Y,\ Y \xrightarrow{S} Z}{X \xrightarrow{ST} Z}$$

and is justified by the integral

$$ST(x, C) =_D \int_Y S(y,C)T(x,dy) \ .$$

The relevant integral calculus and much more is well known and can be found, for example, in Neveu (1965). One novel point we will explore in some detail here is the observation that integral calculus and measure theory, though essential in actual computation, are not needed when the interest focuses on conceptualization. Integrals operate on a microlevel as it were, which for the most part need not be made explicit. The fact that instead of integrals one could consider finite or countable sums is obvious. Transitions resemble in some sense number-valued relations, and so they can also be thought of as logic style devices whose composition is based on an "existential quantification," averaging out the middle space Y. Our basic idea is that spaces and transitions may be regarded as primitives and their simple axiomatic properties (couched in compositions, products, and localizations) capture much of what one needs and does in classical decision theory. The abstract theory that results has many useful models which specialize not only to stochastic kernels (transition and conditional probabilities) and stochastic matrices (when the spaces are countable), but also to less obvious notions, such as subtransitions, transition capacities (upper and lower measures), quantum transitions, Markov transitions, and the like. In this paper I will mainly look at quantum applications.

The two-sorted "probabilistic language" $\mathbb{P}$ we are about to investigate will use X, Y, Z, ... as individual variables ranging over the class of (intended measurable) <u>spaces</u> and S, T, P, ... will be the variables for <u>transitions</u>. Each space X carries its own identity <u>transition</u> $I_X : X \longrightarrow X$ which corresponds to the usual characteristic function:

$$I_X(x,A) = 1 \text{ if } x \in A \text{ and } 0 \text{ otherwise,}$$

for all x in X and A in $\underline{B}_X$.

Given a chain of transitions

$$X \xrightarrow{P} Y \xrightarrow{T} Z \xrightarrow{S} W,$$

by way of an assumption we have

$$\text{(i)} \quad S(TP) = (ST)P;$$

$$\text{(ii)} \quad TI_Y = T;$$

$$\text{(iii)} \quad I_Z T = T.$$

These axioms are confirmed by the integral-based definition of composition. The intended statistical interpretation of the diagram

$$X \xrightarrow{T} Y \xrightarrow{S} Z$$

is as follows: X is the <u>state</u> (phase, parameter, hypothesis) space, Y is the <u>measurement</u> (observation) space, and Z is the <u>decision</u> (action, inference) space. Furthermore, T is thought of as the <u>conditional distribution</u> in Y, governing the observation procedure of states in X, and S is the <u>randomized decision</u> (strategy) to be found via optimization. The partial ordering (to be discussed later) refers to the <u>informativeness</u> of T′s and the <u>risk</u> of S′s. Localization characterizes the variety of <u>restrictions</u> concerning sample spaces, event algebras, and probabilities.

Among the distinguished spaces I will include the <u>unit space</u> $\mathbb{1}$, corresponding to the singleton measurable space, endowed with the trivial two-element Boolean algebra. Transitions of the form

$$\mathbb{1} \xrightarrow{P} X$$

play the role of probabilities, living on X. For example, if X is the space representing the outcomes of a physical die, then $\mathbb{1}$ represents the rolling mechanism, involving two events: "The die is rolled" and "Nothing happens," where the first event fits the sole outcome. The probability P is the law, governing the die′s probabilistic behavior. Here the random variable aspect is trivial. Recall that set-based functions with singleton domains $\mathbb{1} \xrightarrow{f} X$ are identifiable with the membership $x \in X$, where $f(\cdot) = x$. So, $X \xrightarrow{T} Y$ may be thought of as a generalized (parameterized) probabilistic element $\mathbb{1} \xrightarrow{T(x)} Y$, one for each x.

Another simple constant space is 2, representing the doubleton sample space, typical not only in coin-tossing experiments and probabilistic logic, but also in hypothesis testing and representation of events by means of 2-valued random variables. For many, many reasons it is good to accept the real line $\mathbb{R}$ as another extra constant space. It carries the Borel sets in a natural way. Many other important spaces will enter the scene via constructions.

The composition of the Bayesian diagram

$$\mathbb{1} \xrightarrow{P} X \xrightarrow{T} Y$$

is the conceptual content of the well-known example of two urns in a cascade: A ball is drawn from the first urn (described by X) and placed in the second urn (described by Y) with transition (conditional) probability T. Knowing the probability law P for the first urn, we get the final probability of drawing a particular ball from the second urn simply by composition TP. The reader remembers that what comes out computationally is a weighted sum of conditional probabilities--a very special instance of the integral

$$TP(B) = \int_X T(x,B)P(dx) .$$

Cascade composition of transitions is not enough. One needs a way of combining transitions in parallel. It is the job of the <u>product</u> whose formal rule is

$$\frac{X \xrightarrow{T} Y, \quad Z \xrightarrow{S} W}{X \otimes Z \xrightarrow{T \otimes S} Y \otimes W}$$

and which is justified by the product definition

$$[T \otimes S](\langle x,z \rangle, B \times D) = T(x,B)S(z,D)$$

with the obvious domains for variables.

The reader is certainly familiar with the product space $X \otimes Y$ from elementary probability calculus, where the underlying sample space is the Cartesian product $X \times Y$ and the Boolean algebra of events is

$$\underline{B}_{X \times Y} = \underline{B}_X \otimes \underline{B}_Y ,$$

the smallest Boolean algebra containing all products $A \times B$ with A in $\underline{B}_X$ and B in $\underline{B}_Y$. The idea can be exploited in a number of ways. One has <u>powers</u> $X^{\otimes n+1} = X^{\otimes n} \otimes X$, infinite and even continuous products, etc. Independent repetitions of experiments, Markovian processes, and the like need these constructions.

On a theoretical level, products suggest a number of important derived notions. For example, the <u>diagonal product</u>

$$\frac{X \xrightarrow{T} Y, \quad X \xrightarrow{S} Z}{X \xrightarrow{T \Delta S} Y \otimes Z}$$

which is defined for Cartesian products of events by

$$[T \Delta S](x, B \times C) = T(x,B)S(x,C)$$

and then extended to arbitrary events by addition, is an important construction. Before handling the remaining derived constructs, I will pause to explain the status of measurable maps.

Granted a transition $X \xrightarrow{T} Y$, it may happen that, for every x, T(x) is actually a Dirac measure concentrated at one point y of the codomain. In this case, T is in 1:1 correspondence with a <u>measurable map</u> $t : X \longrightarrow Y$ such that $y = t(x)$. Looked at conversely, given a measurable map $f : X \longrightarrow Y$, we get a unique transition $f^{\Delta} : X \longrightarrow Y$ by putting

$$f^{\Delta}(x,B) = 1 \text{ if } f(x) \in B \text{ and } 0 \text{ otherwise.}$$

Obviously, $f^{\Delta}(x)$ is a Dirac measure on Y at point f(x) and $f(\bullet,B)$ is the characteristic function (indicator) of the inverse image $f^{-1}(B)$. The identity measurable map $1_X : X \longrightarrow X$ gives $1_X^{\Delta} = I_X$.

If $\mathbb{P}[X,Y]$ denotes the set of all transitions from X to Y, then the Dirac transitions single out the set

$$\mathbb{R}[X,Y] \subset \mathbb{P}[X,Y]$$

of all measurable maps from X to Y. One can accept this as a definition of measurable maps.

In what follows, the variables f, g, h, ... will range over the class of measurable maps. It is well known that measurable maps compose along usual lines. One even has a product $f \times g : X \otimes Z \longrightarrow Y \otimes W$ of $f : X \longrightarrow Y$ and $g : Z \longrightarrow W$, defined coordinatewise: $[f \times g](x,z) = \langle f(x), g(z) \rangle$.

A measurable map $f : X \longrightarrow Y$ is called a <u>measurable isomorphism</u> if and only if it is 1:1 and onto and its inverse is also measurable. In this case we write $X \cong Y$.

We put

$$\mathbb{R}(X) =_D \mathbb{R}[X,\mathbb{R}]$$

for the space of <u>real random variables</u>. A suitable choice of a Boolean algebra will convert it into a new space of the language $\mathbb{P}$. The endless profusion of linear and other structures on $\mathbb{R}(x)$ is well known.

Since measurable maps are special kinds of transitions, it is expected that they will interact reasonably with transitions. Indeed, that is the case. Given a chain

$$\mathbb{1} \xrightarrow{P} X \xrightarrow{f} Y ,$$

by the <u>direct image of</u> P <u>under</u> f we mean the probability $P_f : \mathbb{1} \longrightarrow Y$, defined by

$$P_f(b) = P(f^{-1}(B))$$

for all B in $\underline{B}_Y$. When $Y = \mathbb{R}$, P_f is usually called the <u>distribution law</u> of the random variable f.

The idea works equally well for transitions in general. Given a diagram

$$X \xrightarrow{f} Y \xrightarrow{T} Z \xrightarrow{g} W ,$$

a transition $T_g : Y \longrightarrow W$ is called the <u>direct image of</u> T <u>under</u> g if and only if

$$[T_g](y) = [T(y)]_g$$

for all y in Y. (Please keep in mind that T(y) is a probability measure!) On the other hand, $T^f : X \longrightarrow Z$ is called the <u>inverse image of</u> T <u>under</u> f if and only if

$$[T^f](x) = T(f(x))$$

for all x in X.

The composition of

$$X \xrightarrow{T} Y \xrightarrow{f} \mathbb{R}$$

is defined as before, fitting the integral

$$[fT](x) = \int_Y f(y)T(x,dy).$$

Here transitions transport random variables from one domain into another in a contravariant fashion. Using the obvious domains, one can quickly verify the associativity

$$g(TS) = (gT)S$$

and a number of related propeties. When $X = \mathbb{1}$, the composition reduces to the usual P-<u>expectation</u> of the random variable $f : Y \longrightarrow \mathbb{R}$:

$$E_P[f] = \int_Y f(y)P(dy) \ .$$

The restriction of this integral to events causes the random variables to carry measures into measures.

I will leave this subject with the following simple theorem which summarizes much of what we know about variable substitution in integration calculus. The clauses may be viewed as axioms for direct and inverse image constructions.

THEOREM 1.

(1) $T_{gf} = (T_f)_g$,

$$\text{if } X \xrightarrow{T} Y \xrightarrow{f} Z \xrightarrow{g} W;$$

(2) $T^{gf} = (T^g)^f$,

$$\text{if } X \xrightarrow{f} Y \xrightarrow{g} Z \xrightarrow{T} W;$$

(3) $T_{1_Y} = T^{1_X} = T$,

$$\text{if } X \xrightarrow{T} Y;$$

(4) $(T_g)^f = (T^f)_g$,

$$\text{if } X \xrightarrow{f} Y \xrightarrow{T} Z \xrightarrow{g} W;$$

(5) $(ST)_g = (S_g)T$,

$$\text{if } X \xrightarrow{T} Y \xrightarrow{S} Z \xrightarrow{g} W;$$

(6) $(ST)^f = S(T^f)$,

$$\text{if } X \xrightarrow{f} Y \xrightarrow{T} Z \xrightarrow{S} W;$$

(7) $S(T_f) = (S^f)T$,

$$\text{if } X \xrightarrow{T} Y \xrightarrow{f} Z \xrightarrow{S} W;$$

(8) $I_f T = T_f$,

$$\text{if } X \xrightarrow{T} Y \xrightarrow{f} Z.$$

Let $\mathbb{M}(X)$ be the (convex) space of all probability measures on X. For every event A of X we define a map

$$A^* : \mathbb{M}(X) \longrightarrow \mathbb{R}$$

by putting $A^*(P) = P(A)$ for all measures P. The smallest Boolean algebra of subsets of $\mathbb{M}(X)$ under which all A^* are measurable, converts $\mathbb{M}(X)$ into a legitimate space of $\mathbb{P}$. Actually, we can simply define

$$\mathbb{M}(X) = \mathbb{P}[\mathbb{1}, X]$$

as it was alluded to earlier. This construction may be iterated and used for defining <u>random probabilities</u>. Indeed, suppose R(Z) is a fixed space of random variables. Then by a <u>random probability</u> on space X we mean a map

$$Q : \underline{B}_X \longrightarrow \mathbb{R}(Z)$$

such that the following clauses hold almost everywhere:

(i) $Q(X) = 1$;

(ii) $Q(A) \geq 0$;

(iii) $Q(A \cup B) = Q(A) + Q(B)$, if $A \cap B = \emptyset$,

where A and B are in $\underline{B}_X$.

If Z is equipped with a probability, then one is able to compute probabilities of events concerning other probabilities. In quantum mechanics we will use this construction rather often. The idea generalizes to <u>random transitions</u>

$$X \xrightarrow{T} Y/Z$$

where T(x) is a random probability on Y for all x, taking its values in the space of random variables $\mathbb{R}(Z)$.

A probability $\mathbb{1} \xrightarrow{P} Z$ and a random probability $\mathbb{1} \xrightarrow{Q} Y/Z$ lead to the composition $\mathbb{1} \xrightarrow{QP} Y$, as seen from

$$[QP](B) = \int_Z [Q(B)](z)P(dz) ,$$

where B is in $\underline{B}_Y$. The measure QP is actually the probability image of P under Q and it agrees with the earlier definition of composition. Since random probabilities are in a correspondence with transitions, for the remaining development we shall ignore them.

The space $\mathbb{M}(X)$ is rich in structure. It is closed under <u>convex mixing</u>:

$$P,Q \in M(X) \Longrightarrow P \underset{a}{+} Q \in \mathbb{M}(X) ,$$

where

$$[P \underset{a}{+} Q](A) = aP(A) + (1 - a)Q(A)$$

with $0 \leq a \leq 1$. The properties of mixing are known in the theory of mixture spaces. Furthermore, it has an <u>orthogonality</u> device

$$P \perp Q \text{ iff } \exists A[P(A) = 0 \ \& \ Q(A) = 1] ,$$

satisfying the usual symmetry and monotonicity:

$$(P \perp R \ \& \ Q \perp R) \Longrightarrow P \underset{a}{+} Q \perp R .$$

It also has a preorder, namely, <u>absolute continuity</u>:

$$P << Q \text{ iff } \forall A[Q(A) = 0 \Longrightarrow P(A) = 0] ,$$

enjoying the monotonicity

$$P << Q \Longrightarrow TP << TQ, \text{ if } \mathbb{1} \underset{Q}{\overset{P}{\rightrightarrows}} X \xrightarrow{T} Y .$$

Every transition $X \xrightarrow{T} Y$ induces a convex mixture and orthogonality preserving measurable map $\mathbb{M}(X) \xrightarrow{T_*} \mathbb{M}(Y)$, defined by $T_*(P) = TP$ for all $P : \mathbb{1} \longrightarrow X$, whose job is to carry X-probabilities into Y-probabilities. Clearly, $(ST)_* = S_*T_*$ and $(I_X)_* = 1_{\mathbb{M}(X)}$.

Dually, we have a contravariant measurable map $\mathbb{R}(Y) \xrightarrow{T^*} \mathbb{R}(X)$, defined by $T^*(g) = gT$ for all $g : Y \longrightarrow \mathbb{R}$. It carries Y-random variables into X-random variables. Naturally, $(ST)^* = T^*S^*$ and $(I_X)^* = 1_{\mathbb{R}(X)}$. The familiar integral

$$\int_X T^*(g)dP = \int_Y gdT_*(P)$$

is the microversion of the associativity $(gT)P = g(TP)$, mentioned earlier.

Every transition $X \xrightarrow{T} Y$ gives a measurable map $X \xrightarrow{T(\cdot)} \mathbb{M}(Y)$ and vice versa. The result is that we can initiate an alternative Grundlagen for decision theory, starting with measurable maps as primitives.

The Dirac map $\Delta_X : X \longrightarrow \mathbb{M}(X)$ defined by

$$[\Delta_X(x)](A) = 1 \quad \text{if} \quad x \in A \quad \text{and} \quad 0 \quad \text{otherwise}$$

leads to the conversions $f^\Delta = \Delta_Y f$ and $T^f = Tf^\Delta$ for every $X \xrightarrow{f} Y \xrightarrow{T} Z$.

Getting back to the diagonal product, we can define

$$T \Delta S =_D (T \otimes S)^\delta ,$$

as seen from the diagram

$$X \xrightarrow{\delta} X \otimes X \xrightarrow{T \otimes S} Y \otimes Z ,$$

where δ is the measurable diagonal map, defined by $\delta(x) = \langle x,x \rangle$. When $X = \mathbb{1}$, we get the usual definition of product probabilities.

A much more important notion is the cascade (joint) product

$$\frac{X \xrightarrow{T} Y,\ Y \xrightarrow{S} Z}{X \xrightarrow{S*T} Y \otimes Z}$$

where

$$S*T =_D (I \Delta S)T = (I \otimes S)^{\delta} T .$$

Given

$$\mathbb{1} \xrightarrow{P} X \xrightarrow{T} Y ,$$

in the language of integrals we have

$$[T*P](A \times B) = \int_A T(x,B)P(dx) ,$$

which supports the connection

$$[S*T](x) = S*T(x) .$$

The above Bayesian diagram with the joint product is the conceptual content of the ways in which one gets joint probabilities from transitions (conditionals) and marginals. It is a generalization of the Bayesian idea $P(A \cap B) = P(B/A)P(A)$. The definition, of course, works for every event C in $X \otimes Y$:

$$[T*P](C) = \int_X T(x, i_x^{-1}(C))P(dx)$$

where

$$\mathbb{1} \xrightarrow{P} X \xrightarrow{T} Y \xrightarrow{i_x} X \otimes Y \text{ and } i_x(y) = \langle x,y \rangle .$$

In applied probability one is frequently faced with a situation in which several experiments are given, out of which one is selected randomly on the basis of an auxiliary experiment. The conceptualization leads to <u>mixing</u>. Given an auxiliary probability $\mathbb{1} \xrightarrow{P} Z$ and a transition $X \xrightarrow{T} Y$, we say that $Z \otimes X \xrightarrow{S} Y$ is a P-<u>mixture</u> of Y if and only if $S(I_X \otimes P) = T$, i.e.,

$$\int_Z S(\langle z,x \rangle, B)P(dz) = T(x,B) .$$

Hence we have a new rule

$$\frac{Z \otimes X \xrightarrow{\ S\ } Y}{X \xrightarrow{\ S^+\ } \int_Z Y dP}$$

conveying the following idea: We are given a family (measurable field) of transitions $X \xrightarrow{S_z} Y(z)$, indexed by Z such that $Z \otimes X \xrightarrow{S} Y$, defined by $S(\langle z,x\rangle,B) = S_z(x,B)$, is a transition. The auxiliary probability $P : \mathbb{1} \longrightarrow Z$ picks out one transition S_z. This amounts to saying that S^+ is a mixture experiment in the (Hilbert integral) space $\int_Z YdP$. When $Z \cong 2$ with outcomes a and b, the Hilbert integral collapses into a <u>sum</u> $Y_a \oplus Y_b$, whose underlying sample space is the disjoint (separated) sum $Y_a + Y_b$ and whose algebra

$$\underline{B}_{Y_a + Y_b} = \underline{B}_{Y_a} \oplus \underline{B}_{Y_b}$$

is the smallest Boolean algebra containing the disjoint sums $A + B$ with A in $\underline{B}_{Y_a}$ and B in $\underline{B}_{Y_b}$. Now the experiment is carried out with probability $P(a)$ in Y_a and with probability $P(b)$ in Y_b.

<u>Conditionalization</u> is a by-product of localization and transition. Every nonempty subspace B of the space X leads to a <u>state restriction</u> $X|B$ whose underlying sample space is B, and whose algebra of events is the <u>trace</u>

$$\underline{B}_X|B = \{A \cap B \,|\, A \in \underline{B}_X\} .$$

Besides state restrictions we have <u>event restrictions</u>: Given a space X and a Boolean subalgebra $\underline{G}$ of $\underline{B}_X$, by $X|\underline{G}$ we denote the new space whose states are the same as before but the algebra of events is $\underline{G}$. (The possible combinations $(X|B)|\underline{G}$ and $(X|\underline{G})|B$ should not detain us.)

Given $\mathbb{1} \xrightarrow{\ P\ } X$ and a subalgebra $\underline{G}$ of $\underline{B}_X$, a transition

$$X \xrightarrow{\ P|\underline{G}\ } X|\underline{G}$$

is called a <u>conditional</u> <u>probability</u> <u>of</u> P, <u>given</u> $\underline{G}$, if and only if

(i) $[P|\underline{G}](x) \in \mathbb{M}(X)$ for all x in X;

(ii) $P(A|\underline{G}) =_D [P|\underline{G}](\bullet,A) \in \mathbb{R}(X|\underline{G})$ for all A in $\underline{B}_X$;

(iii) $P(A \cap B) = [P|\underline{G}*P](A \times B)$ for all A in $\underline{B}_X$ and $B \in \underline{G}$.

That is,

$$P(A \cap B) = \int_B [P|\underline{G}](x,A)P(dx) ,$$

where the left-hand side is actually the restriction $P|B : \mathbb{1} \longrightarrow X|B$ at event A.

The transition $P|\underline{G}$ is unique modulo some P-equivalence and its random aspect disappears when $\underline{G}$ is generated by a countable measurable partition (B_i). Indeed, then we have the classical case:

$$P(A/B_i) =_D \frac{P(A \cap B_i)}{P(B_i)} = [P|\underline{G}](x,A) = [P(A|\underline{G})](x)$$

for x in B_i, where the conditional comes out as a normalization of a restriction probability. The familiar substitution rule

$$\int_X [P|\underline{G}](\bullet,A)P_f(dx) = \int_X (P|\underline{G})^f(\bullet,A)P(dx)$$

and many other similar rules will fall out from the formalism. Conditionals may be extended to transitions:

$$[T|\underline{G}](x) = T(x)|\underline{G} .$$

Finishing up images, products, and restrictions, I will present the following theorem:

THEOREM 2

(1) $X \otimes \mathbb{1} \cong \mathbb{1} \otimes X \cong X$;

(2) $X \otimes Y \cong Y \otimes X$;

(3) $X \otimes (Y \otimes Z) \cong (X \otimes Y) \otimes Z$;

(4) $\mathbb{M}(X \otimes Y) \cong \mathbb{M}(X) \otimes \mathbb{M}(Y)$;

(5) $\mathbb{R}(X \otimes Y) \cong \mathbb{R}(X) \otimes \mathbb{R}(Y)$;

(6) $\mathbb{M}(1) \cong \mathbb{1}$ and $\mathbb{R}(\mathbb{1}) \cong \mathbb{R}$;

(7) Let $\mathbb{1} \xrightarrow{P} X \xrightarrow{f} Y \xrightarrow{T} Z \xrightarrow{g} W$.

Then

(i) $T_g{*}P_f = (T^f{*}P)_{f\times g}$;

(ii) $T_gP_f = (T^fP)_g$.

(8) Let $\mathbb{1} \xrightarrow{P} X \xrightarrow{T} Y \xrightarrow{S} Z$.

Then

(i) $(S{*}T)P = S{*}TP$;

(ii) $(S{*}T){*}P = S^{p_2}{*}(T{*}P)$, if $X \otimes Y \xrightarrow{p_2} Y$;

(iii) $STP = S^{p_2}(T{*}P)$;

(iv) $ST{*}P = ((S{*}T){*}P)^{p_{23}}$, if

$X \otimes (Y \otimes Z) \xrightarrow{p_{23}} Y \otimes Z$;

(9) $I_X{*}P = P_\delta$, if $\mathbb{1} \xrightarrow{P} X \xrightarrow{\delta} X \otimes X$.

(10) $(T{*}P)_{p_1} = T(\bullet,Y)P$, if $\mathbb{1} \xrightarrow{P} X \xrightarrow{T} Y$

and $X \otimes Y \xrightarrow{p_1} X$.

The proof is straightforward, based on the definitions of the concepts involved.

The last concept we need to discuss is the ordering of P[X,Y]. It is a partial ordering (modulo equivalence) with properties

(i) $T \preccurlyeq S \Longrightarrow UT \preccurlyeq US$;

(ii) $T \preccurlyeq S \Longrightarrow TU \preccurlyeq SU$;

(iii) $T \preccurlyeq S \Longrightarrow T|\underline{G} \preccurlyeq S|\underline{G}$;

(iv) $(T \preccurlyeq S \ \&\ P \preccurlyeq Q) \Longrightarrow T \otimes P \preccurlyeq S \otimes Q$,

where the domains should be obvious.

It is time to present the principal idea of decision. Given a Bayesian scheme

$$\mathbb{1} \xrightarrow{P} X \xrightarrow{T} Y \xrightarrow{S} Z ,$$

we think of P as the <u>initial</u> (prior) probability and we think of T as the <u>experiment</u> (observational transition), and S is the <u>strategy</u> to be selected on the basis of prior knowledge and observation, given <u>loss</u> (cost) h. Based on the conceptual facilities we have, the scheme can be rewritten as follows:

$$\mathbb{1} \xrightarrow{P} X \xrightarrow{ST*I} X \otimes Z \xrightarrow{h} \mathbb{R} .$$

The composition $h(I_X \otimes S)$ is the <u>average loss</u> and

$$R_S =_D h(ST*I_X) : X \longrightarrow \mathbb{R}$$

is the <u>risk</u>, dependent on the strategy S. The preference on decisions $S \preccurlyeq S'$ is often interpreted as $R_S \leq R_{S'}$ (the natural order on random variables, modulo the P-probability of measure zero). Hence the main point is to choose S in such a way that the risk is <u>minimal</u>. Of course, other representation requirements are possible. The set of decisions $\mathbb{P}[Y,Z]$ is too big and has to be cut down to the subset of <u>admissible decisions</u>: those to which there is nothing strictly preferable. Now in this set, a <u>Bayesian strategy</u> may be chosen according to the rule

$$S \preccurlyeq S' \quad \text{iff} \quad R_S P \leq R_{S'} P .$$

If Z = 2, then the scheme is applicable to <u>hypothesis testing</u>. The outcomes in 2 are z_0 = accept the hypothesis $x \in H$ and z_1 = reject the hypothesis $x \in H$, where H in $\underline{B}_X$ is the hypothesis event we want to sort out. As far as the loss is concerned, we may put $h(x,z_i) = 0$ if the decision is correct; otherwise, positive. We need not look for a randomized strategy S but maybe only for a nonrandomized Dirac strategy $s : Y \longrightarrow 2$. Here, to find s is equivalent to specifying the <u>critical region</u> $\{y \mid s(y) = z_1\} \subseteq Y$.

If $Z = \mathbb{R}$, then we may try to <u>estimate</u> the value of a random variable $f : X \longrightarrow \mathbb{R}$. The loss could be a quadratic function $h(x,z) = h'(x)(f(x) - z)^2$. Here the strategy $s : Y \longrightarrow \mathbb{R}$ is called the <u>estimator</u>, estimating f(x) in terms of s(y).

If $Z = X$, then the Bayesian scheme

$$\mathbb{1} \longrightarrow X \underset{S}{\overset{T}{\rightleftarrows}} Y$$

leads to the equation $S^*TP = T^*P$ which (in a Radon-Nikodym fashion) may be written as a "derivative"

$$S = \frac{T^*P}{TP}$$

giving an explicit formula for the unknown strategy. Many of these decision schemes will reappear in quantum decision without any conceptual change.

We compare not only decisions but also experiments. For example, we may use again the risk and put

$$T \preccurlyeq T' \Longleftrightarrow \forall S, S' \; [R_{S'} \leq R_S]$$

to say that T' is at least as <u>informative</u> as T. The relevant scheme is

$$\begin{array}{ccc} X & \xrightarrow{T} & Y \\ {\scriptstyle T'}\downarrow & & \downarrow{\scriptstyle S} \\ Y' & \xrightarrow{S'} & Z \end{array} .$$

Certain thresholds in the inequality concerning risks will lead to semiorders which often turn out to be more realistic. Other definitions are possible, using entropy measures, Blackwellian almost-commutativity, etc.

I presented here an abstract theory of decision, admitting standard interpretation, based on composition, product, localization, and partial ordering of transitions. In conditionalization one important question concerns <u>independence</u>; we will have no use for it here.

III. QUANTUM DECISION THEORY

The basic conceptual scenario of quantum mechanics typically begins with a Hilbert space H over one of the division rings whose orthomodular lattice of closed subspaces (or projection operators) $\underline{Q}_H$ is viewed as the correct algebra of events. Its algebra of self-adjoint operators plays the role of the space of <u>observables</u>. By way of a free bonus, the inner product induces probability values for events and expectation values for observables. In view of the availability of a rather large number of deep representation theorems, much of this can be said abstractly, for example, without ever mentioning Hilbert spaces.

The situation is grossly analogous to the classical triad: phase space of states, Boolean algebra of events, and the space of random variables. The details, however, are radically different. States are probabilities on $\underline{Q}_H$, events are not always distributive and not even modular, observables do not commute in general, and joint probabilities are often lost. Mathematicians isolated the notion of <u>noncommutativity</u> as the key concept responsible for the difference. This author favors the viewpoint that the difference between classical and quantum reasoning lies in <u>partialness</u> or localization. Quantum concepts (such as quantum lattices, convex sets of quantum probabilities) appear less strange when viewed as some sort of generalized manifolds which are coverable by classical entities. For example, in Domotor (1974) it is explained how abstract orthomodular lattices may be represented by atlases of covering Boolean algebras. The essence of the local viewpoint lies in a clever reformulation of the classical notion of random variable (which is useless in quantum contexts). The inverse image $f^{-1} : \mathbb{B}_R \longrightarrow \underline{B}_X$ of every random variable $f : X \longrightarrow \mathbb{R}$ is a Boolean homomorphism. (In fact, there is a 1:1 correspondence between random variables and Boolean homomorphisms, modulo the ideal of Borel sets of Lebesgue measure zero.)

By way of a generalization one assumes that <u>quantum observables</u> are point-free quantum lattice homomorphisms $u : \underline{B}_X \longrightarrow \underline{Q}_H$ which (in agreement with our earlier notation) will be denoted by

$$H \xrightarrow{\ u\ } X .$$

The virtues of this symbolism will be appreciated later.

In what follows, we shall extend the earlier frame $\mathbb{P}$ by adding to the old spaces a class of Hilbert spaces H, K, ... The new spaces have <u>products</u> $H \otimes K$, <u>sums</u> $H \oplus K$, and localizations. The event algebras $\underline{Q}_{H \times K} = \underline{Q}_H \otimes \underline{Q}_K$ and $\underline{Q}_{H+K} = \underline{Q}_H \oplus \underline{Q}_K$ may be used freely to represent events of coupled quantum systems. New spaces bring with themselves new (quantum) observables and sigma homomorphisms u, v, ... We also have quantum probabilities

$$\mathbb{1} \xrightarrow{W} H$$

which formally remain the same as classical probabilities (in the sense that abstractly the Kolmogorovian axioms still hold), except that the events are now elements of the lattice $\underline{Q}_H$. The consequences are severe. For example, the classical implication $(W(E) = 1 \ \& \ W(E') = 1) \Longrightarrow W(E \wedge E') = 1$ and its dual $(W(E) = 0 \ \& \ W(E') = 0) \Longrightarrow W(E \vee E') = 0$ are false in general.

By way of a direct image, the chain

$$\mathbb{1} \xrightarrow{W} H \xrightarrow{u} R$$

gives the probability law $W_u : \mathbb{1} \longrightarrow \mathbb{R}$ of the observable u. Hence,

$$W_u(B) = W(u(B))$$

is the probability that the observable u takes its values in the Borel set B.

One Bayesian scheme to look at is

$$\mathbf{1} \xrightarrow{P} X \xrightarrow{W} H \xrightarrow{u} Z ,$$

where P is the initial (prior) probability of the parameter space X, W is a parameterized family of quantum probabilities on H, and u is a quantum observable to be optimized. In many cases, $u : H \longrightarrow Z$ is interpreted as a method of measurement. That is, Z is the space of possible states (scale points) of an observing system whose algebra of events (possible readings) is $\underline{B}_Z$. The quantum homomorphism u associates with each reading (event) of the observing device a unique quantum event in $\underline{Q}_H$, concerning the observed system.

In accordance with classical reasoning, we allocate a loss h and get a modified diagram

$$X \xrightarrow{W_u * I_X} X \otimes Z \xrightarrow{h} \mathbb{R} .$$

So that the <u>risk</u> of the observable (strategy) u is

$$R_u = h(W_u * I_X) \ .$$

All we have done is to find again the old integral

$$R_u(x) = \int_Z h(x,z)W(x,u(dz))$$

with quantum probabilities W(x). From this moment on, classical decision theory takes over with all the available resources. In particular, we isolate <u>admissible</u> strategies and then decide about the correct u on the basis of $R_u P$, where P is the prior probability. What should be kept in mind is that the decision concerns the <u>best observable</u> u which (it is hoped) will give the <u>largest amount</u> of information about the <u>unknown state</u> W, under given loss h. In classical systems this issue is meaningless, since optimal observables always exist (e.g., position and momentum).

Bayesian inference may be modified in a number of ways. For example, if we replace Z by $\mathbb{M}(Z)$, we get a variant of <u>randomized strategies</u>. One associates probability distributions (and not just events) with the particular quantum events.

What I said so far is conceptually correct but computationally uninteresting. Decision-theoretic reasoning is overemphasized at the cost of quantum reasoning. To get things right, we must pass to operators and then perhaps to matrices. Real-valued quantum observables on H (as discussed before) are in 1:1 correspondence with Hermitean self-adjoint operators acting on H and the states are in 1:1 correspondence with nonnegative trace class operators of trace 1, living again on H. Specifically, Gleason's famous theorem says that every probability $\mathbb{1} \xrightarrow{W} H$ is of the form

$$W(E) = \mathrm{Tr}[W' \cdot \hat{E}] \ ,$$

where E is an event in $\underline{Q}_H$ and $\hat{E}$ is its projection operator, and W′ is a nonnegative trace class density operator with Tr(W′) = 1. Trace Tr is a linear functional on the algebra of (bounded) Hermitean operators $\mathbb{R}(H)$, giving the probability values.

In what follows, $\mathbb{1} \xrightarrow{W} H$ will denote such a density operator (or even its matrix representation). The 1:1 correspondence makes this abuse of notation harmless. Similar translations are available for other self-adjoint operators, in particular, events may be thought of as projections.

Along the lines of random probabilities, we shall consider <u>operator-valued probabilities</u>, defined as maps

$$U : \underline{B}_Z \longrightarrow \mathbb{R}(H)$$

such that

(i) $U(Z) = 1_H$ (identity operator);

(ii) $U(A) \geq 0$;

(iii) $U(A \cup B) = U(A) + U(B)$, if $A \cap B = \emptyset$;

(iv) $\operatorname{Lim}_n \sum_{i \leq n} U(A_i) = 1_H$, if (A_i) is a countable measurable partition of Z.

To appreciate properly the significance of this concept, the reader should recall the notion of <u>qualitative quantum experiment</u> (a countable partition of the unit element of $\underline{Q}_H$). It happens to be the quantum variant of the classical 'qualitative random variable' (countable measurable partition of a sample space). Given such an experiment (E_i) over H, its quantum events E_j are the <u>possible outcomes</u>. Now similarly, $U(A_j)$ may be thought of as an outcome of the 'experiment U' over a partition (A_i). In important special cases, $U(A_j)$ is actually a projection operator (hence a quantum event). Unlike quantum events in an experiment (E_i), the operators $U(A_j)$ need not be compatible (commutative).

Another way to assess the importance of operator probabilities is to recall that in the standard accounts of quantum theory an observable is represented by a self-adjoint operator $\dot{U}$ on H which (in view of the spectral theorem) enjoys the representation

$$\dot{U} = \int_R aU(da)$$

where $U : B_{\mathbb{R}} \longrightarrow \mathbb{R}(H)$ is a projection-valued probability. This fact suggests the possibility of redefining observables as operator probabilities. Every operator probability U determines a self-adjoint operator $\dot{U}$, but not necessarily the other way around.

The idea of measurement of $\dot{U}$, then, is really a vastly generalized scheme of a coin-tossing experiment:

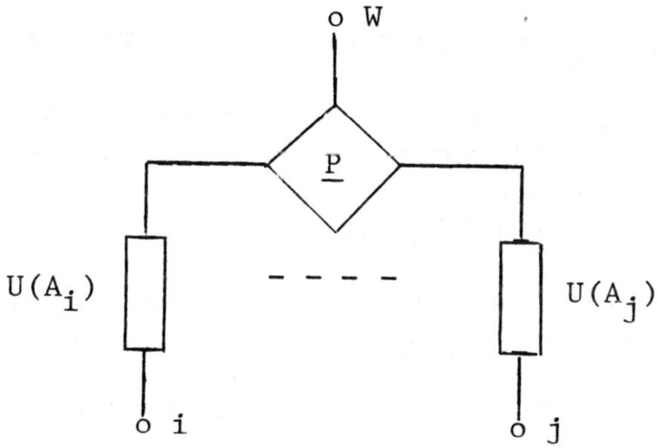

An instrument accepts a state W, measures $\dot{U}$, and emits an output state conditional on the value observed. The exit at node i occurs with probability $Tr[W \cdot U(A_i)]$.

The set of all operator-valued probabilities is denoted by $\mathbb{M}_H(Z)$. This space is an analogue of $\mathbb{M}(Z)$, discussed earlier, and is equally rich in structure. It is closed under <u>convex mixing</u>

$$U, V \in \mathbb{M}_H(Z) \Longrightarrow U \underset{a}{+} V \in M_H(Z) ,$$

with the same meaning as in the case of classical probabilities. It is preordered by <u>dominance</u> (absolute continuity) relation <<, defined again as before. Furthermore, it is endowed with two new relations: <u>orthogonality</u> and <u>compatibility</u> (commutativity).

For further discussion of quantum decision and quantum experiments, it is important to single out three interesting subsets of $\mathbb{M}_H(Z)$:

(a) <u>Orthogonal observables (operator probabilities)</u>

$$\overset{\perp}{M}_H(Z) =_D \{U \mid \forall A, B [A \cap B = \emptyset \Longrightarrow U(A) \cdot U(B) = 0]\} .$$

In this case, inevitably, the operators $U(A)$ are the projections, as one can quickly check from the spectral decomposition of U. The coordinatewise orthogonality $U(A) \perp V(A)$ for all A in $\underline{B}_Z$ defines the orthogonality $U \perp V$, mentioned before. The elements of $\mathbb{M}_H(Z)$ may be identified with qualitative quantum experiments.

(b) <u>Commutative observables</u> (<u>operator probabilities</u>)

$$\mathbb{M}_H^{\|}(Z) =_D \{U \mid \forall A,B[U(A)\cdot U(B) = U(B)\cdot U(A)]\} .$$

The coordinatewise commutativity of $U(A)$ and $V(A)$ for all A in $\underline{B}_Z$ defines the commutativity $U \| V$, suggested earlier.

(c) <u>Extreme</u> (<u>boundary</u>) <u>observables</u>

Since $\mathbb{M}_H(Z)$ is a convex set, one may expect to be able to isolate its <u>extreme</u> points:

$$\mathbb{M}_H^{\partial}(Z) = \{U \mid U = V \underset{a}{+} V' \Longrightarrow U = V = V'\} .$$

These concepts will help organize the machinery of quantum decision.

In analogy with random probabilities, we should write $U : \mathbb{1} \longrightarrow Z/H$ for every operator probability U on Z. Its composition with density $W : \mathbb{1} \longrightarrow H$ gives the probability image of W under U, namely $UW : \mathbb{1} \longrightarrow Z$, where

$$[UW](A) = \mathrm{Tr}[U(A)\cdot W]$$

is the probability that the measurement of U in Z yields a value in $A \in \underline{B}_Z$, when the observed system (represented by H) is in a (mixed) state W. We know that for pure states $w : \mathbb{1} \longrightarrow H$ we have the usual

$$[Uw](A) = \mathrm{Tr}[U(A)w] = \langle U(A)w,w\rangle = \|U(A)w\|^2 .$$

For reasons of transparency and conformity with the classical situation, we shall continue to write

$$\mathbb{1} \xrightarrow{\;W\;} H \xrightarrow{\;U\;} Z$$

to represent the measurement of U when the system is in state W. Also, we will retain the old notation W_U

instead of UW, to emphasize that it is the direct image of W under U. The presence of operators makes substitution look like a product. An important point to bear in mind from now on is that operator probabilities are actually <u>generalized quantum observables</u> U whose measurement outcomes U(A) are again operators--in many cases just projections (quantum events).

Now a typical quantum Bayesian scheme is of the form

$$\mathbb{1} \xrightarrow{P} X \xrightarrow{W} H \xrightarrow{U} Z$$

which together with loss h gives

$$\mathbb{1} \xrightarrow{P} X \xrightarrow{W_U * I_X} X \otimes Z \xrightarrow{h} R$$

and thus

$$R_U = h(W_U * I_X)$$

is the risk operator to be minimized with respect to observable U. Again, the corresponding integral

$$R_U(x) = \int_Z h(x,z)\mathrm{Tr}[W(x)\cdot U(dz)]$$

gives the risk of guessing the true (unknown) state W(x) of the system based on measuring the observable U. As pointed out, not all observables are equally informative and unlike in the classical case, in quantum mechanics there is no universally optimal measurement! The point here is to find the most informative observable U, regarding W, given a loss and other constraints.

The partial ordering $U \preccurlyeq V$, known in classical decision theory, agrees again with the risk operator comparison $R_U \leq R_V$.

The Bayesian tactic is to choose an observable U from a suitable class of operators, based on the minimal value of the affine functional $R_U P$.

In concrete situations often the trace functional commutes with the averaging integrals and the computation for minimization may be simplified. For example, if we put

$$K(z) = (W \,\Delta\, hi_z)P$$

where $i_Z : X \longrightarrow X \otimes Z$ is the usual section, then we can minimize

$$R_U = \mathrm{Tr}[\int_Z K(z)\cdot U(dz)] .$$

The following theorem summarizes the basic facts concerning optimal measurement of observables.

THEOREM 3 (Naimark-Holevo).

(1) <u>Every orthogonal observable</u> $H \xrightarrow{U} Z$ <u>is an extreme point of</u> $\mathbb{M}_H(Z)$:

$$\overset{\perp}{\mathbb{M}}_H(Z) \subset \overset{\partial}{\mathbb{M}}_H(Z) .$$

(2) <u>Every locally commuting observable</u> $H \xrightarrow{U} Z$ (i.e., $\exists A,B[A \neq B \ \& \ U(A)\cdot U(B) = U(B)\cdot U(A)]$ <u>in</u> $\overset{\partial}{\mathbb{M}}_H(Z)$ <u>is also in</u> $\overset{\perp}{\mathbb{M}}_H(Z)$.

(3) <u>Every commutative observable</u> U <u>in</u> $\overset{\parallel}{\mathbb{M}}_H(Z)$ <u>has a unique counterpart</u> U' in $\overset{\perp}{\mathbb{M}}_H(\mathbb{R}(Z))$. <u>That is, randomization reduces commutative observables to orthogonal observables</u>:

$$U(A) = \int_{\mathbb{M}(Z)} P(A)U'(dP) .$$

(4) <u>Given a measurement</u>

$$\mathbb{1} \xrightarrow{W} H \xrightarrow{U} Z ,$$

<u>then there exists another</u> (nonunique) <u>measurement</u>

$$\mathbb{1} \xrightarrow{W \otimes W_0} H \otimes H_0 \xrightarrow{U'} Z$$

<u>such that</u>

(i) $U' \in \mathbb{M}_{H \otimes H_0}(Z)$;

(ii) $W_U = (W \otimes W_0)_{U'}$.

That is, every quantum measurement of an arbitrary observable U is statistically equivalent to a measurement of an orthogonal observable U′ on some extension of the initial system. (The initial system is coupled with an auxiliary system, ancilla, which remains in a fixed state W_0.)

This theorem is proved in Holevo (1972), using a related result of Naimark. One of the most important facts of quantum measurement is stated in Clause 4. It is central to measuring noncommutative observables.

To see what is going on, I will present an example, based on a sketch in Holevo (1972).

Let

$$\mathbb{1} \xrightarrow{\;P\;} X \xrightarrow{\;W\;} H \xrightarrow{\;U\;} Z$$

be a quantum Bayesian scheme, where $X = \{x_1, x_2, x_3\}$ (the set of possible parameters) has a uniform prior $P(x_i) = 1/3$, and $Z = \{z_1, z_2, z_3\}$ (the set of possible decisions) has a loss $h(x_i, z_j) = 1 - d_{ij}$ (here d is the Kronecker function). Suppose that H is a two-dimensional Euclidean space with densities (hypotheses to be guessed) $W(x_i) = E_i$ $(i = 1,2,3)$. We assume that E_i are projections on three plane vectors with angles $2\pi/3$.

Our problem is to find the best observable U in the sense that it will be optimal for learning about the hypotheses E_i. Decision theory suggests to compute the operator

$$K(z_j) = \sum_{i=1}^{3} h(x_i, z_j) P(x_i) W(x_i) = \sum_{i=1}^{3} (1 = d_{ij})\ 1/3\ E_i$$

$$= 1/3\ (E_{j_1} + E_{j_2}) = 1/2\ 1_H - 1/3\ E_j \ .$$

[Remember that in view of our assumption we have $2/3\ (E_1 + E_2 + E_3) = 1_H$.]

The risk

$$R_U P = \mathrm{Tr}\Big[\sum_{j=1}^{3} K(z_j) \cdot U(z_j)\Big]$$

is minimal (and equal to 1/3) for observable $U(z_j) = 2/3\ E_j$. This observable is not orthogonal, although it is extremal. Holevo points out that if we had taken an orthogonal U, the minimal risk would have been much greater. Thus in order to get the best possible information about the unknown state, we should look for a more general observable than those which give orthogonal projections as outcomes.

IV. MEASUREMENT OF NONCOMMUTING OBSERVABLES

Given a quantum lattice $\underline{Q}_H$ and two quantum events E and E′ from it, the compatibility (simultaneous observability) is defined by $E \| E'$ if and only if E and E′ belong to the same subalgebra of $\underline{Q}_H$. This relation is not an equivalence, but still it has a number of well-understood properties.

The notion of compatibility is extended quickly to observables coordinatewise:

$$u \| v =_D \forall A, B\, [u(A) \| v(B)], \quad \text{if} \quad H \underset{v}{\overset{u}{\rightrightarrows}} \mathbb{R} .$$

For observables with $u \| v$ the joint observable

$$H \xrightarrow{u;v} \mathbb{R} \otimes \mathbb{R}$$

exists and so does the joint probability $W_{u;v}$. It is the direct image of W under u;v.

In the operator setting, two observables are compatible if and only if they commute. Hence, if $U \| V$, then the measurement scheme

$$1 \xrightarrow{W} H \underset{V}{\overset{U}{\rightrightarrows}} Z$$

gives a joint observable and then a joint distribution $W_{U;V} : \mathbb{1} \longrightarrow Z \otimes Z$, where

$$W_{U;V}(B \times C) = \mathrm{Tr}[W \cdot U(B); W \cdot U(C)]$$

is the probability that a simultaneous measurement of U and V in $Z \otimes Z$ yields a value in $B \times C$, when the system is in state W.

Now what if U and V do not commute? Then in view of the Naimark-Holevo Theorem 3(4), one can find two <u>commuting</u> observables U′ and V′ such that

$$\mathbb{1} \xrightarrow{W \otimes W_0} H \otimes H_0 \xrightarrow{U';V'} Z \otimes Z$$

will lead to a joint measurement for some W_0 and H_0. Moreover, U′ and V′ can be selected optimally as estimators of U and V, respectively. One may look at the quadratic loss $(U \otimes 1 - U')^2 + (V \otimes 1 - V')^2$ and then proceed to compute $R_{U';V'}$.

By way of an example, let p and q be the position and momentum, respectively, of a harmonic oscillator. Suppose that the state is Gaussian with generating function $W(e^{i(xp+yq)}) = e^{(-\sigma^2/2)(x^2+y^2)}$. Hence $W(p^2) = W(q^2) = \sigma^2 = Tr(p^2 \cdot W)$ and Heisenberg′s estimate would be $W(p^2) \cdot W(q^2) \geq 1/4$ and so $\sigma^2 \geq 1/2$.

Surprisingly, the Naimark-Holevo theorem offers a much better estimate for the joint measurement of p and q. All we have to do is to select a new pair of commuting observables

$$p' = c(p \otimes 1 + 1 \otimes p_o)$$

$$q' = c(q \otimes 1 + 1 \otimes q_o)$$

with some constants p_o, q_o, and c, and we may take W_0 to be the vacuum state.

After selecting the quadratic error we may put $K(a,b) = (p - a1)W(p - a1) + (q - b1)W(q - b1)$ and then proceed to minimize the functional $R_{p';q'} = Tr[\iint K(a,b) \cdot [p';q'](da,db)]$. Holevo indicates that in the case of dispersion $\sigma^2 = 1/2$, we have $W[(p' - p)^2] \cdot W[(q' - q)^2] = 1/16$, which is considerably less than the estimate obtained from the Heisenbergian formula for uncertainty.

Helstrom and Kennedy (1974) have applied Holevo′s technique for estimating simultaneous measurements of noncommuting observables in quantum receivers.

References

Domotor, Z. The probability structure of quantum-mechanical systems. *Synthese*, 1974, *29*, 155-185.

Helstrom, C. W., & Kennedy, R. S. Noncommuting observables in quantum detection and estimation theory. *IEEE Transactions on Information Theory*, 1974, *IT-20*, 16-24.

Holevo, A. S. Statistical problems in quantum physics. In G. Maruyama & Y. V. Prokhorov (Eds.), *Proceedings of the Second Japan-USSR Symposium on Probability Theory*. (Springer Lecture Notes in Mathematics, No. 330.) New York: Springer, 1972.

Neveu, J. *Mathematical foundations of the calculus of probability*. San Francisco: Holden-Day, 1965.

BIBLIOGRAPHY ON QUANTUM LOGIC

Janet Beehner
Stanford University

Books

[1] *Actes du Congrès International de Philosophie des Sciences*. Paris: Librairie Scientifique, 1936.

[2] *Akten des XIV internationalen Kongresses für Philosophie* (Vol. 3). Vienna: Herder, 1969.

[3] *American Mathematical Society proceedings of Symposia in Pure Mathematics* (Vol. 2). Providence, R.I.: American Mathematical Society, 1961.

[4] *Applications scientifiques de la logique mathématique* (P. Destouches-Février, Ed.). (*Actes du 2ème Colloque International de Logique Mathématique*.) Paris: Gauthier-Villars, 1952.

[5] *Applied probability* (L. A. MacColl, Ed.). (*Proceedings of Symposia in Applied Mathematics*, Vol. 7.) New York: McGraw-Hill, 1957.

[6] *Lecture notes in physics* (Berlin). *Springer-Verlag Lecture Notes*, 1974, *24*. (A set of lectures presented at the NATO Advanced Study Institute on Quantum Mechanics and Ordered Linear Spaces, 1973.)

[7] *Proceedings of the Conference on Convexity and Combinatorial Geometry*. Norman: University of Oklahoma Press, 1971.

[8] Abbott, J. C. (Ed.). *Trends in lattice theory*. New York: Van Nostrand, 1970.

Studies in the Foundations of Quantum Mechanics, pp. 223-261

[9] Ackermann, R. *An introduction to many-valued logics*. London: Routledge & Kegan Paul, 1967.

[10] Addison, J., Henkin, L., & Tarski, A. (Eds.). *The theory of models*. Amsterdam: North-Holland, 1965.

[11] Audi, M. *The interpretation of quantum mechanics*. Chicago: University of Chicago Press, 1973.

[12] Bar-Hillel, Y. *Logic, methodology and the philosophy of science*. Amsterdam: North-Holland, 1965.

[13] Bates, D. R. (Ed.). *Quantum theory* (Vol. 3). New York: Academic Press, 1962.

[14] Belinfante, F. J. *A survey of hidden-variable theories*. New York: Pergamon Press, 1973.

[15] Bharucha-Reid, A. T. (Ed.). *Probabilistic methods in applied mathematics* (Vol. 2). New York: Academic Press, 1970.

[16] Bohm, D. *Quantum theory*. Englewood Cliffs, N.J.: Prentice-Hall, 1951.

[17] Bub, J. *The interpretation of quantum mechanics*. (*University of Western Ontario series in philosophy of science*, Vol. 3.) Dordrecht: Reidel, 1974.

[18] Bunge, M. *Foundations of physics*. Berlin: Springer-Verlag, 1967.

[19] Butts, R. E., & Hintikka, J. (Eds.). *Foundational problems in the special sciences*. (*University of Western Ontario series in philosophy of science*, Vol. 10.) Boston: Reidel, 1977.

[20] Castell, L., Drieschner, M., & von Weizsäcker, C. F. (Eds.). *Quantum theory and the structures of time and space*. Munich: Carl Hanser Verlag, 1975.

[21] Cohen, R. S., Hooker, C. A., Michalos, A. C., & van Evra, J. W. (Eds.). *PSA 1974*. (*Boston studies in the philosophy of science*, Vol. 32.) Dordrecht: Reidel, 1976.

[22] Cohen, R. S., & Wartofsky, M. W. (Eds.). Proceedings of the Boston Colloquium for the Philosophy of Science 1966/1968. (Boston studies in the philosophy of science, Vol. 5.) Dordrecht: Reidel, 1969.

[23] Cohen, R. S., & Wartofsky, M. W. (Eds.). Logical and epistemological studies in contemporary physics. (Boston studies in the philosophy of science, Vol. 13.) Dordrecht: Reidel, 1974.

[24] Colodny, R. (Ed.). Frontiers of science and philosophy. (University of Pittsburgh series in philosophy of science, Vol. 1.) Pittsburgh: University of Pittsburgh Press, 1962.

[25] Colodny, R. (Ed.). Paradigms and paradoxes: The philosophical challenge of the quantum domain. (University of Pittsburgh series in philosophy of science, Vol. 5.) Pittsburgh: University of Pittsburgh Press, 1972.

[26] Crawley, P., & Dilworth, R. P. Algebraic theory of lattices. Englewood Cliffs, N.J.: Prentice-Hall, 1973.

[27] D'Espagnat, B. Conceptual foundations of quantum mechanics (2nd ed.). Redding, Mass.: Benjamin, Addison-Wesley, 1976.

[28] D'Espagnat, B. (Ed.). Foundations of quantum mechanics: Proceedings of the International School of Theoretical Physics `Enrico Fermi' (Course 49). New York: Academic Press, 1972.

[29] Destouches-Février, P. La structure des théories physiques. Paris: Presse Universitaire de France, 1951.

[30] Dockx, S., & Bernays, P. Information and prediction in science. New York: Academic Press, 1965.

[31] Donnellan, T. Lattice theory. Oxford: Pergamon Press, 1968.

[32] Dummett, M. *Frege's philosophy of language*. London: Duckworth, 1973. (Chapter 17.)

[33] Feyerabend, P. K., & Maxwell, G. (Eds.). *Mind, matter, and method*. (*Essays in philosophy and science in honor of Herbert Feigl*.) Minneapolis: University of Minnesota Press, 1966.

[34] Flato, M., Maric, Z., Milojevic, A., Sternheimer, D., & Vigier, J. P. *Quantum mechanics, determinism, causality and particles*. Dordrecht: Reidel, 1976.

[35] Freund, P. C. O., Goebel, C. J., & Nambu, Y. (Eds.). *Quanta*. Chicago: University of Chicago Press, 1970.

[36] Gilbert, R. P., & Newton, R. G. (Eds.). *Analytic methods in mathematical physics*. New York: Gordon & Breach, 1970.

[37] Haack, S. *Deviant logic: Some philosophical issues*. Cambridge, England: Cambridge University Press, 1974.

[38] Harper, W. L., & Hooker, C. A. *Foundations of probability theory, statistical inference and statistical theories of science*. (*University of Western Ontario series in philosophy of science*, Vol. 3.) Dordrecht: Reidel, 1976.

[39] Henkin, L., Suppes, P., & Tarski, A. (Eds.). *The axiomatic method with special reference to geometry and physics*. (*Studies in logic and the foundations of mathematics*.) Amsterdam: North-Holland, 1959.

[40] Heisenberg, W. *Physics and philosophy*. New York: Harper & Row, 1966.

[41] Hooker, C. A. (Ed.). *Contemporary research in the foundations and philosophy of quantum theory*. (*University of Western Ontario series in philosophy of science*, Vol. 2.) Dordrecht: Reidel, 1973.

[42] Hooker, C. A. (Ed.). *The logico-algebraic approach to quantum mechanics*. Dordrecht: Reidel, 1975.

[43] Jammer, M. The philosophy of quantum mechanics. New York: Wiley, 1974. (Chapter 8.)

[44] Jauch, J. M. Foundations of quantum mechanics. Reading, Mass.: Addison-Wesley, 1968. (MR 36 #1151.)

[45] Klibansky, R. (Ed.). Contemporary philosophy: A survey (Vol. 2). Firenze, Italy: La Nuova Italia Editrice, 1968.

[46] Kronig, R., & Weisskopf, V. F. (Eds.). Collected scientific papers (Vol. 2). New York: Interscience, 1964.

[47] Lambert, K. (Ed.). The logical way of doing things. New Haven, Conn.: Yale University Press, 1969.

[48] Ley, H., & Lother, R. (Eds.). Mikrokosmos, makrokosmos (Vol. 2). Berlin: Akademie-Verlag, 1967.

[49] Mackey, G. The mathematical foundations of quantum mechanics. New York: Benjamin, 1963. (MR 27 #5501.)

[50] Maeda, F., & Maeda, S. Theory of symmetric lattices. Berlin: Springer-Verlag, 1970.

[51] Mehra, J. (Ed.). The physicist's conception of nature. Dordrecht: Reidel, 1973.

[52] Menne, A. (Ed.). Logico-philosophical studies. Dordrecht: Reidel, 1962.

[53] Mittelstaedt, P. Philosophical problems of modern physics. (Boston studies in the philosophy of science, Vol. 18.) Dordrecht: Reidel, 1976.

[54] Mittelstaedt, P. Philosophische Probleme der modernen Physik. Mannheim: Bibliographisches Institut, 1963. (English translation, [53].)

[55] Neumann, J. von. The mathematical foundations of quantum mechanics (R. Beyer, trans.). Princeton, N.J.: Princeton University Press, 1955. (MR 16, p. 654.)

[56] Neumann, J. von. Mathematische Grundlagen der Quantenmechanik. Berlin: Springer, 1932. (MR 5, p. 165.)

[57] Nikodym, O. M. The mathematical apparatus for quantum-theories; Based on the theory of Boolean lattices. (Die Grundlehren der mathematischen Wissenschaften, Band 129.) New York: Springer-Verlag, 1966.

[58] Nilson, D. Quantum logic, quantum probability, and quantum measurement: A philosophical perspective on the quantum theory. Unpublished doctoral dissertation, Indiana University, 1972.

[59] Pearce, G., & Maynard, P. (Eds.). Conceptual change. Dordrecht: Reidel, 1973.

[60] Piron, C. Foundations of quantum physics. (Mathematical Physics Monographs, No. 19.) New York: Benjamin, 1975.

[61] Przelecki, M. The logic of empirical theories. London: Routledge & Kegan Paul, 1969.

[62] Reichenbach, H. Philosophic foundations of quantum mechanics. Los Angeles: University of California Press, 1944. (Edited portion of secs. 29-37 reprinted as "Three valued logic and the interpretation of quantum mechanics" in [42].)

[63] Rescher, N. Many-valued logic. New York: McGraw-Hill, 1969.

[64] Rosser, J. B., & Turquette, A. R. Many valued logics. Amsterdam: North-Holland, 1952.

[65] Rutherford, D. E. Introduction to lattice theory. Edinburgh: Oliver & Boyd, 1968.

[66] Scheibe, E. The logical analysis of quantum mechanics. (J. B. Sykes, trans.) New York: Pergamon Press, 1973.

[67] Stenius, E. Critical essays. (Acta Philosophica Fennica, Vol. 25.) Amsterdam: North-Holland, 1972.

[68] Strauss, M. Modern physics and its philosophy. Dordrecht: Reidel, 1972.

[69] Suppes, P. (Ed.). Logic and probability in quantum mechanics. Dordrecht: Reidel, 1976.

[70] Varadarajan, V. S. Geometry of quantum theory (2 vols.). Princeton, N.J.: Van Nostrand, 1968.

[71] Watanabe, S. Knowing and guessing: A quantitative study of inference and information. New York: Wiley, 1969.

[72] Weingartner, D. (Ed.). Description analytizität und Existenz. Munich: Pustet, 1966.

[73] Yourgrau, W., & van der Merwe, A. (Eds.). Perspectives in quantum theory. Cambridge, Mass.: M.I.T. Press, 1971.

[74] Zinov′ev, A. A. Philosophical problems of many-valued logic (Rev. ed.; G. Küng & D. D. Comey, Eds. and trans.). Dordrecht: Reidel, 1963.

Articles

[75] Adjukiewicz, K. Logic and experience. Synthese, 1949/1951, 8, 289-299.

[76] Albertson, J. Von Neumann′s hidden parameter proof. American Journal of Physics, 1961, 29, 478-484. (MR 25 #5705.)

[77] Alda, V. On hidden variables. Aplikace Matematiky, 1972, 17, 53-55. (MR 44 #6219.)

[78] Arens, R. Invariant sublogics as a way from scalar to many-component wave equations. Journal of Mathematics and Mechanics, 1966, 15, 349-371. (MR 32 #8699.)

[79] Aron, J. C. About hidden variables. Progress of Theoretical Physics, 1969, 42, 715-727. (MR 40 #6858.)

[80] Banks, P. On the philosophical interpretation of logic: An Aristotelian dialogue. In [52].

[81] Basson, A. H. Logic and fact. Analysis, 1947/1948, 8, 81-87.

[82] Bell, J. S. On the problem of hidden variables in quantum mechanics. Reviews of Modern Physics, 1966, 38, 447-452. (MR 34 #8735.)

[83] Beltrametti, E. G., & Cassinelli, G. On the logic of quantum mechanics. Zeitschrift für Naturforschung, 1973, 8a, 1516-1530.

[84] Beltrametti, E. G., & Cassinelli, G. Quantum mechanics and p-adic numbers. Foundations of Physics, 1972, 2, 1-7. (MR 48 #10328.)

[85] Benedetti, A., & Teppati, G. The decision problem for mathematical structures of quantum theory. Lettere al Nuovo Cimento, 1971, 11(2), 695-696. (MR 50 #9193.)

[86] Beth, E. W. Towards an up-to-date philosophy of natural science. Methodos, 1949, 1, 178-185.

[87] Birkhoff, G. Lattices in applied mathematics. In American Mathematical Society Proceedings of Symposium in Pure Mathematics, 1961, 2, 155-184.

[88] Birkhoff, G., & von Neumann, J. The logic of quantum mechanics. Annals of Mathematics, 1936, 37, 823-843. (Also in [42].)

[89] Bohm, D. Hidden variables in the quantum theory. In [13]. (MR 23 #B2832.)

[90] Bohm, D. A suggested interpretation of the quantum theory in terms of "hidden" variables, I and II. Physical Review, 1952, 85, 166-179.

[91] Bohm, D. A refutation of the proof by Jauch and Piron that hidden variables can be excluded in quantum mechanics. Reviews of Modern Physics, 1966, 38, 470-475. (MR 34 #8737.)

[92] Britton, K. Empirical foundations for logic. Analysis, 1933/1934, 2, 37-42.

[93] Bub, J. Hidden variables and quantum logic. Unpublished manuscript, University of Western Ontario, no date.

[94] Bub, J. On the completeness of quantum mechanics. In [42].

[95] Bub, J. Under the spell of Bohr (Review of T. Bastin (Ed.), Quantum theory and beyond). British Journal for the Philosophy of Science, 1973, 24, 78-90.

[96] Bub, J. What is philosophically interesting about quantum mechanics? In [19].

[97] Bub, J. The statistics of non-Boolean event structures. In [38].

[98] Bub, J. Von Neumann's projection postulate as a probability conditionalization rule in quantum mechanics. Journal of Philosophical Logic, 1977, 6, 381-390.

[99] Bub, J., & Demopoulos, W. Critical studies of Paradigms and paradoxes. Philosophia, 1976, 6, 333-344.

[100] Bub, J., & Demopoulos, W. The interpretation of quantum mechanics. In [22].

[101] Bugajska, K. Representation theorem for quantum logic. International Journal of Theoretical Physics, 1974, 9, 93.

[102] Bugajska, K., & Bugajski, S. Hidden variables and 2-dimensional Hilbert space. Annales de l'Institut Henri Poincaré, 1972, 16, 93.

[103] Bugajska, K., & Bugajski, S. The lattice structure of quantum logics. Annales de l'Institut Henri Poincaré, 1973, 19, 333-340. (MR 50 #12009.)

[104] Bugajska, K., & Bugajski, S. On the axioms of quantum mechanics. Bulletin de l'Académie Polonaise des Sciences, 1972, 20, 231-234. (MR 46 #6725.)

[105] Bugajska, K., & Bugajski, S. The projection postulate in quantum logic. Bulletin de l'Académie Polonaise des Sciences, 1973, 21, 871-877. (MR 48 #7796.)

[106] Bugajski, S. Probability implication in the logic of classical and quantum mechanics. Journal of Philosophical Logic, 1978, 7, 95-106.

[107] Bunge, M. Philosophy and physics. In [45].

[108] Bunge, M. Quanta and philosophy. Proceedings of the Seventh Inter-American Congress of Philosophy, 1967, 281-296.

[109] Butrick, R. Discussion: Putnam's revolution. Philosophy of Science, 1971, 38, 290-292.

[110] Catlin, D. Spectral theory in quantum logics. International Journal of Theoretical Physics, 1968, 1, 285-298.

[111] Church, A. On the law of excluded middle. Bulletin of the American Mathematical Society, 1928, 34, 75-78.

[112] Church, A. Review of G. Birkhoff and J. von Neumann, The logic of quantum mechanics. Journal of Symbolic Logic, 1937, 2, 44-45.

[113] Church, A. A set of postulates for the foundation of logic. Annals of Mathematics, 1932, 33, 346-366.

[114] Cirelli, R., & Cotta-Ramusino, P. On the isomorphism of a "quantum logic" with the logic of the projections in a Hilbert space. International Journal of Theoretical Physics, 1973, 8, 11-29.

[115] Cirelli, R., & Gallone, F. Algebra of observables and quantum logic. Annales de l'Institut Henri Poincaré, 1973, 19, 297-331.

[116] Clark, I. D. An axiomatiziation of quantum logic. Journal of Symbolic Logic, 1973, 38, 389-392. (MR 50 #3785.)

[117] Cole, E. A. B. Perception and operation in the definition of observable. International Journal of Theoretical Physics, 1973, 8, 155-170.

[118] Cook, T. A. Geometry of infinite quantum logics--Preliminary report. Notices of the American Mathematical Society, 1975, 22A, 338.

[119] Cook, T. A. Hahn-Jordan decomposition theorem in infinite quantum logics. Notices of the American Mathematical Society, 1975, 22A, 83.

[120] Cook, T. A., & Ruttiman, G. T. Symmetries on quantum logics--Preliminary report. Notices of the American Mathematical Society, 1975, 22A, 663-664.

[121] Costa de Beauregard, O. Extension d'une théorie de J. de Neumann au cas des projecteurs non commutables. Comptes Rendus Hebdomadaires des Séances de l'Académie des Sciences, 1945, 21, 230-231.

[122] Curry, H. B. The relation of logic to science. In [30].

[123] Czelakowski, J. Logics based on partial Boolean σ-algebras, I. Studia Logica, 1974, 33, 371-396. (MR 50 #15622.)

[124] Czelakowski, J. Logics based on partial Boolean σ-algebras, II. Studia Logica, 1975, 34, 69-86. (MR 51 #7979.)

[125] Dalla Chiarra, M. L. A general approach to non-distributive logics. Studia Logica, 1976, 35, 139-162.

[126] Dalla Chiarra, M. L. Logical self-reference, set theoretical paradoxes and the measurement problem in quantum mechanics. _Journal of Philosophical Logic_, 1977, _6_, 331-347.

[127] Dalla Chiarra, M. L. Quantum logic and physical modalities. _Journal of Philosophical Logic_, 1977, _6_, 391-404.

[128] Dalla Chiarra, M. L., & Toraldo di Francia, G. A logical analysis of physical theories. _Rivista del Nuovo Cimento_, 1973, _3_, 1-20.

[129] Deliyann, P. C. Generalized hidden variable theorem. _Journal of Mathematical Physics_, 1971, _13_, 1013.

[130] Deliyann, P. C. Superposition of states and structure of quantum logics. _Journal of Mathematical Physics_, 1976, _17_, 248-254.

[131] Deliyann, P. C. Vector-space models of abstract quantum logics. _Journal of Mathematical Physics_, 1973, _14_, 249-253.

[132] Demopoulos, W. Completeness and realism in quantum mechanics. In [19].

[133] Demopoulos, W. Critical notice on Hooker's _Contemporary research in the foundations and philosophy of quantum theory_. _Synthese_, 1976, _33_, 489-504.

[134] Demopoulos, W. The possibility structure of physical systems. In [38].

[135] Demopoulos, W. _What is the logical interpretation of quantum mechanics_? Unpublished manuscript, University of Western Ontario, no date.

[136] Denecki, H.-M. Quantum logic of quantifiers. _Journal of Philosophical Logic_, 1977, _6_, 405-414.

[137] D'Espagnat, B. Quantum logic and non-separability. In [51].

[138] Destouches-Février, P. Logique et théories physiques. *Synthese*, 1948/1949, *7*, 400-410.

[139] Destouches-Février, P. Applications des logiques modales en physique quantique. *Theoria*, 1952, *1*, 167-169.

[140] Destouches-Février, P. La logique de propositions expérimentales. In [4].

[141] Dishkant, H. The first-order predicate calculus based on the logic of quantum mechanics. *Reports on Mathematical Logic*, 1974, 9-17.

[142] Dishkant, H. Imbedding of the quantum logic in the modal system of Brouwer. *Journal of Symbolic Logic*, 1977, *42*, 321-328.

[143] Dishkant, H. Semantics of the minimal logic of quantum mechanics. *Studia Logica*, 1972, *30*, 23-32.

[144] Dombrowski, H. D., & Horneffer, K. Der Begriff des physikalischen Systems in mathematischer Sicht. *Nachrichten der Akademie der Wissenschaften Göttingen*, 1964, *2*, 67-100. (*MR* 31 #2978.)

[145] Dorling, J. Three reviews of works on quantum logic. *British Journal for the Philosophy of Science*, 1976, *27*, 295-302.

[146] Duckenfield, C. J. A continuous geometry as a mathematical model for quantum mechanics. *Commentationes Mathematicae Universitatis Carolinae*, 1969, *10*, 217-236. (*MR* 40 #3807.)

[147] Eckman, J., & Zabey, P. C. Impossibility of quantum mechanics in a Hilbert space over a finite field. *Helvetica Physica Acta*, 1969, *62*, 420-424. (*MR* 39 #7904.)

[148] Emch, G., & Jauch, J. M. Structures logiques et mathématiques en physique quantique. *Dialectica*, 1965, *19*, 259-279.

[149] Emch, G., & Piron, C. Symmetry in quantum theory. Journal of Mathematical Physics, 1963, 4, 469-473. (MR 26 #5856.)

[150] Erwin, E. Quantum logic and the status of classical logic. Unpublished manuscript; a version read at the American Philosophical Association, Pacific Division, spring 1974.

[151] Erwin, E. Are the notions "a priori truth" and "necessary truth" extensionally equivalent? Canadian Journal of Philosophy, 1974, 3, 591-602.

[152] Falk, G. Axiomatik als Methode physikalischer Theorienbildung. Zeitschrift für Physik, 1951, 130, 51-68. (MR 13, p. 96.)

[153] Farber, M. Logical systems and the principles of logic. Philosophy of Science, 1942, 9, 40-54.

[154] Fáy, G. A phenomenological foundation of quantum logic. Acta Physica Academiae Scientiarum Hungaricae, 1970, 29, 27-33. (MR 45 #1486.)

[155] Février, P. Logical structure of physical theories. In [39]. (MR 22 #518.)

[156] Février, P. Les relations d'incertitude de Heisenberg et la logique. Comptes Rendus Hebdomadaires des Séances de l'Académie des Sciences, 1937, 204, 481-483.

[157] Finch, P. D. On the lattice structure of quantum logic. Bulletin of the Australian Mathematical Society, 1969, 1, 333-340. (MR 41 #4914.)

[158] Finch, P. D. On the structure of quantum logic. Journal of Symbolic Logic, 1969, 34, 275-282. (Reprinted in [42].)

[159] Finch, P. D. Quantum logic as an implication algebra. Bulletin of the Australian Mathematical Society, 1970, 2, 101-106. (MR 41 #4915.)

[160] Fine, A. Logic, probability, and quantum theory. _Philosophy of Science_, 1968, _35_, 101-111.

[161] Fine, A. Probability and the interpretation of quantum mechanics. _British Journal for the Philosophy of Science_, 1973, _24_, 1-37.

[162] Fine, A. Some conceptual problems of quantum theory. In [21] and [25].

[163] Finkelstein, D. Classical and quantum probability and set theory. In [38].

[164] Finkelstein, D. The Leibniz project. _Journal of Philosophical Logic_, 1977, _6_, 425-439.

[165] Finkelstein, D. The logic of quantum physics. _Transactions of the New York Academy of Sciences_, 1962/1963, _25_, 621-637. (_MR_ 27 #5503.)

[166] Finkelstein, D. Matter, space and logic. In [21].

[167] Finkelstein, D. The physics of logic. In [25].

[168] Finkelstein, D. A process conception of nature. In [51].

[169] Finkelstein, D. Space-time code. _Physical Review_, 1969, _184_, 1261-1271.

[170] Finkelstein, D. Space-time code, II. _Physical Review_, 1972, _D5_, 320-328.

[171] Finkelstein, D. Space-time code, III. _Physical Review_, 1972, _D5_, 2922-2931.

[172] Finkelstein, D. Space-time code, IV. _Physical Review_, 1974, _D9_, 2219-2231.

[173] Finkelstein, D., Frye, G., & Suskind, L. Space-time code, V. _Physical Review_, 1974, _D9_, 2231-2236.

[174] Fisk, M. The logic of either-or. _Notre Dame Journal of Formal Logic_, 1965, _6_, 39-50.

[175] Foulis, D. J. Baer*-semigroups. Proceedings of the American Mathematical Society, 1960, 11, 648-654.

[176] Foulis, D. J., & Randall, C. H. Empirical logic and quantum mechanics. Synthese, 1974, 29, 81-111. (Reprinted in [69].)

[177] Fraïssé, R. Essai sur la logique de l'indeterminisme et la ramification de l'espace-temps. In [69].

[178] Fraassen, B. C. van. Bub on the completeness of quantum mechanics. Unpublished manuscript, University of Toronto, no date.

[179] Fraassen, B. C. van. The formal representation of physical quantities. In [21].

[180] Fraassen, B. C. van. Hidden variables in conditional logic. Theoria, 1974, 15, 176-190.

[181] Fraassen, B. C. van. The labyrinth of quantum logics. In [22]. (Reprinted in [42].)

[182] Fraassen, B. C. van. Meaning relations among predicates. Noûs, 1967, 1, 161-179.

[183] Fraassen, B. C. van. Meaning relations and modalities. Noûs, 1969, 3, 155-167.

[184] Fraassen, B. C. van. On the extension of Beth's semantics of physical theories. Philosophy of Science, 1970, 37, 325-339.

[185] Fraassen, B. C. van. Semantic analysis of quantum logic. In [41].

[186] Friedman, M., & Glymour, C. If quanta had logic. Journal of Philosophical Logic, 1972, 1, 16-28.

[187] Fuchs, W. R. Ansätze zu einer Quantenlogik. Theoria, 1964, 30, 137-140.

[188] Gallone, F., & Zecca, A. Quantum logic axioms and the proposition-state structure. _International Journal of Theoretical Physics_, 1973, _8_, 51-63. (_MR_ 50 #15623.)

[189] Gardner, M. R. Is quantum logic really logic? _Philosophy of Science_, 1971, _38_, 508-529.

[190] Gardner, M. R. Quantum-theoretical realism: Popper and Einstein v. Kochen and Specker. _British Journal for the Philosophy of Science_, 1972, _23_, 13-23.

[191] Gardner, M. R. Two deviant logics for quantum theory: Bohr and Reichenbach. _British Journal for the Philosophy of Science_, 1972, _23_, 89-109.

[192] Gleason, A. M. Measures on the closed subspaces of a Hilbert space. _Journal of Mathematics and Mechanics_, 1957, _6_, 885-893. (Reprinted in [42].)

[193] Glymour, C. Review of Bub's _Interpretation of quantum mechanics_. _Canadian Journal of Philosophy_, 1976, _6_, 161-175.

[194] Goldblatt, R. I. Semantic analysis of orthologic. _Journal of Philosophical Logic_, 1974, _3_, 19-35.

[195] Golden, S. A formal theory of quantum classification, I. _Nuovo Cimento_, 1957, _5_ (suppl.), 540-567. (_MR_ 20 #665.)

[196] Gonseth, F. La logique en tant que physique de l'objét quelconque. In [1].

[197] Graves, J. C. Review of Cohen and Wartofsky (Eds.), _Boston studies in the philosophy of science, Volume 5_. _British Journal for the Philosophy of Science_, 1973, _24_, 183-190.

[198] Greechie, R. J. Combinatorial quantum logic. In D. C. Kay (Ed.), _Proceedings of the Conference on Convexity and Combinatorial Geometry_. Norman: University of Oklahoma, Dept. of Mathematics, 1971. (_MR_ 49 #1944.)

[199] Greechie, R. J. Orthomodular lattices admitting no states. _Journal of Combinatorial Theory_, 1971, _10_, 119-132.

[200] Greechie, R. J. Some results from the combinatorial approach to quantum logic. _Synthese_, 1974, _29_, 113-127. In [69]. (Reprinted in _Notices of the American Mathematical Society_, 1975, _22A_, 53.)

[201] Greechie, R. J., & Gudder, S. P. Is a quantum logic a logic? _Helvetica Physica Acta_, 1971, _44_, 238-240. (_MR_ 46 #8551.)

[202] Greechie, R. J., & Gudder, S. P. Quantum logics. In [41]. (Reprinted in [42].)

[203] Gudder, S. P. Axiomatic quantum mechanics and generalized probability theory. In [15].

[204] Gudder, S. P. Complete sets of observables and pure states. _Canadian Journal of Mathematics_, 1968, _20_, 1276-1280. (_MR_ 38 #906.)

[205] Gudder, S. P. Coordinate and momentum observables in axiomatic quantum mechanics. _Journal of Mathematical Physics_, 1967, _8_, 1848-1858. (_MR_ 36 #3541.)

[206] Gudder, S. P. Coordinization of orthomodular posets. _Notices of the American Mathematical Society_, 1969, _16_, 190.

[207] Gudder, S. P. Dispersion-free states and the exclusion of hidden variables. _Proceedings of the American Mathematical Society_, 1968, _19_, 319-324. (_MR_ 36 #7383.)

[208] Gudder, S. P. Hidden variable models for quantum mechanics. _Nuovo Cimento B_, 1972, _10_, 518-522. (_MR_ 50 #12012.)

[209] Gudder, S. P. Hidden variables in quantum mechanics reconsidered. _Reviews of Modern Physics_, 1968, _40_. 229.

[210] Gudder, S. P. Joint distributions of observables. _Journal of Mathematics and Mechanics_, 1968/1969, _18_, 325-335. (_MR_ 38 #907.)

[211] Gudder, S. P. A note on proposition observables. _Pacific Journal of Mathematics_, 1969, _28_, 101-104. (_MR_ 38 #6801.)

[212] Gudder, S. P. On hidden-variable theories. _Journal of Mathematical Physics_, 1970, _11_, 431-436. (_MR_ 41 #4916.)

[213] Gudder, S. P. On the quantum logic approach to quantum mechanics. _Communications in Mathematical Physics_, 1969, _12_, 1-15.

[214] Gudder, S. P. Partial algebraic structures associated with orthomodular posets. _Pacific Journal of Mathematics_, 1972, _41_, 717.

[215] Gudder, S. P. Quantum logics, physical space, position observables and symmetry. _Reports on Mathematical Physics_, 1973, _4_, 193-202. (_MR_ 48 #3379.)

[216] Gudder, S. P. A superposition principle in physics. _Journal of Mathematical Physics_, 1970, _11_, 1037-1040. (_MR_ 41 #6478.)

[217] Gudder, S. P. Systems of observables in axiomatic quantum mechanics. _Journal of Mathematical Physics_, 1967, _8_, 2109-2113. (_MR_ 36 #6193.)

[218] Gudder, S. P. Uniqueness and existence properties of bounded observables. _Pacific Journal of Mathematics_, 1966, _19_, 81-93. (_MR_ 34 #1030.)

[219] Gudder, S. P., & Boyce, L. Projective representations of quantum logics. _International Journal of Theoretical Physics_, 1970, _3_, 99-108.

[220] Gudder, S. P., & Boyer, L. A comparison of the Mackey and Segal models for quantum mechanics. _International Journal of Theoretical Physics_, 1970, _3_, 7-21.

[221] Gudder, S. P., & Boyer, L. Projective representations of quantum logics. _International Journal of Theoretical Physics_, 1970, _3_, 99-108.

[222] Gudder, S. P., & Piron, C. Observables and the field in quantum mechanics. _Journal of Mathematical Physics_, 1971, _12_, 1583-1588.

[223] Guenin, M. Axiomatic foundations of quantum theories. _Journal of Mathematical Physics_, 1966, _7_, 271-282.

[224] Gunson, J. On the algebraic structure of quantum mechanics. _Communications in Mathematical Physics_, 1967, _6_, 262-285. (_MR_ 37 #6087.)

[225] Gunson, J. Physical states on quantum logics, I. _Annales de l'Institut Henri Poincaré_, 1972, _17_, 295-311.

[226] Guz, W. Quantum logic and a theorem on commensurability. _Reports on Mathematical Physics_, 1971, _2_, 53-61. (_MR_ 46 #4866.)

[227] Hacking, I. What is logic? _Journal of Philosophy_, 1979, _76_, 285-319.

[228] Hardegree, G. M. The conditional in quantum logic. _Synthese_, 1974, _29_, 63-80. (Reprinted in [69].)

[229] Hardegree, G. M. _Logic and the semantic view of theories_. Unpublished manuscript, University of Massachusetts, Amherst, no date.

[230] Hardegree, G. M. _A note on the conditional in quantum logic_. Paper presented at the 1974 Philosophy of Science Association Meeting, University of Notre Dame, Notre Dame, Indiana, 1974.

[231] Hardegree, G. M. _Quasi-implication and quantum logic_. Unpublished manuscript, University of Massachusetts, Amherst, no date.

[232] Hardegree, G. M. Stalnaker conditionals and quantum logic. *Journal of Philosophical Logic*, 1975, *4*, 399-421.

[233] Heelan, P. A. Complementarity, context dependence, and quantum logic. *Foundations of Physics*, 1970/1971, *1*, 95-110. (*MR* 47 #8014.)

[234] Heelan, P. A. The logic of framework transpositions. *International Philosophical Quarterly*, 1971, *11*, 314-334.

[235] Heelan, P. A. Quantum and classical logic: Their respective roles. *Synthese*, 1970, *21*, 2-33.

[236] Henkin, L. Review of Putnam, "Three-valued logic"; Feyerabend, "Reichenbach's interpretation of quantum mechanics"; and Levi, "Putnam's three truth-values." *Journal of Symbolic Logic*, 1960, *25*, 289-291.

[237] Herman, L., Marsden, E. L., & Piziak, R. Implication connectives in orthomodular lattices. *Notre Dame Journal of Formal Logic*, 1975, *16*, 305-328.

[238] Hockney, D. Conceptual structures. In [59].

[239] Holland, S. S., Jr. The current interest in orthomodular lattices. In [8]. (Reprinted in [42].)

[240] Hooker, C. A. The nature of quantum mechanical reality: Einstein versus Bohr. In [25].

[241] Hübner, K. Über den Begriff der Quantenlogik. *Sprache in Technischen Zeitalter*, 1964, *22*, 925-934.

[242] Husimi, K. Studies on the foundations of quantum mechanics, I. *Proceedings of the Physico-Mathematical Society of Japan*, 1937, *19*, 766-789.

[243] Inglesby, M. Some criticisms of quantum logic. *Helvetica Physica Acta*, 1971, *44*, 299-307. (*MR* 48 #10330.)

[244] Jauch, J. M. Scattering theory in general quantum mechanics. In [36]. (MR 49 #6793.)

[245] Jauch, J. M., & Misra, B. Supersymmetries and essential observables. Helvetica Physica Acta, 1961, 34, 699-709. (MR 25 #943.)

[246] Jauch, J. M., & Piron, C. Can hidden variables be excluded in quantum mechanics? Helvetica Physica Acta, 1963, 36, 827-837. (MR 30 #1742.)

[247] Jauch, J. M., & Piron, C. On the structure of quantal proposition systems. Helvetica Physica Acta, 1969, 42, 842-848. (Reprinted in [42].) (MR 40 #5191.)

[248] Jauch, J. M., & Piron, C. What is "quantum logic"? In [35].

[249] Jeffcott, B. The center of an orthologic. Journal of Symbolic Logic, 1972, 37, 641-645.

[250] Jeffcott, B. Commuting observables in a σ-orthologic. Indiana University Mathematics Journal, 1973/1974, 23, 369-376. (MR 48 #2007.)

[251] Jeffcott, B. Decomposable othologics. Notre Dame Journal of Formal Logic, 1975, 16, 329-338.

[252] Jordan, P. Bemerkungen zur Quantenlogik. Annales de la Faculté des Sci- ences, Université Clermon Ferrand, 1962, 8, 159-166.

[253] Jordan, P. Quantenlogik und das kommutative Gesetz. In [39].

[254] Jordan, P. Zur axiomatischen Begründung der Quantenmechanik. Zeitschrift für Physik, 1952, 133, 21-29. (MR 14, p. 434.)

[255] Jordan, P., von Neumann, J., & Wigner, E. On an algebraic generalization of the quantum mechanical formalism. Annals of Mathematics, 1934, 35, 29-64.

[256] Kägi-Romano, U. Quantum logic and generalized probability theory. Journal of Philosophical Logic, 1977, 6, 455-462.

[257] Kamber, F. The structure of the propositional calculus of a physical theory. Nachrichten der Akademie der Wissenschaften in Göttingen, 1964, 10, 103-124. (Reprinted in [42].) (MR 32 #5045.)

[258] Kochen, S., & Specker, E. P. The calculus of partial propositional functions. In [12]. (Reprinted in [42].)

[259] Kochen, S., & Specker, E. P. Logical structures arising in quantum theory. In [10]. (Reprinted in [42].) (MR 34 #5409.)

[260] Kochen, S., & Specker, E. P. The problem of hidden variables in quantum mechanics. Journal of Mathematics and Mechanics, 1967, 17, 57-87. (Reprinted in [42].) (MR 36 #2363.)

[261] Koopman, B. O. Quantum theory and the foundations of probability. In [5]. (MR 19, p. 610.)

[262] Körner, S. Logic and conceptual change. In [59].

[263] Kotas, J. An axiom system for the modular logic. Studia Logica, 1967, 21, 17-38.

[264] Kotas, J. The modular logic as a calculus of logical schemata. Studia Logica, 1971, 27, 73-79.

[265] Kraus, K. Algebras of observables with continuous representations of symmetry groups. Communications in Mathematical Physics, 1968, 7, 99-111. (MR 36 #6194.)

[266] Kriesel, G. A notion of mechanistic theory. In [69].

[267] Kripke, S. Kripke has lectured extensively on quantum logic and the question "Is logic empirical?" Various personal copies of the lecture notes are often cited. (Unpublished manuscripts, Princeton University.)

[268] Krips, H. P. Quantal quandries. _Australasian Journal of Philosophy_, 1974, _52_, 133-145.

[269] Kronfli, N. S. Abstract scattering theory. _International Journal of Theoretical Physics_, 1969, _2_, 345-349. (_MR_ 43 #4360.)

[270] Kronfli, N. S. Atomicity and determinism in Boolean systems. _International Journal of Theoretical Physics_, 1971, _4_, 141-143. (Reprinted in [42].)

[271] Kronfli, N. S. Integration theory of observables. _International Journal of Theoretical Physics_, 1970, _3_, 199-204. (Reprinted in [42].)

[272] Kronfli, N. S. Probabilistic formulations of classical mechanics. _International Journal of Theoretical Physics_, 1970, _3_, 395-400. (Reprinted in [42].)

[273] Kunsemüller, H. Zur Axiomatik der Quantenlogik. _Philosophia Naturalis_, 1964, _8_, 363-376.

[274] Kuznetsov, B. G. On quantum-relativistic logic. _Soviet Studies in Philosophy_, 1970/1971, _9_, 203-211.

[275] Kuznetsov, B. G., et al. Soviet Symposium on Logic and Quantum Mechanics. _Soviet Studies in Philosophy_, 1970/1971, _9_, 203-236.

[276] Lambert, K. Logical truth and microphysics. In [47].

[277] Landé, A. ψ superposition and quantum periodicity. Physical Review, 1957, 108, 891-893. (MR 20 #6284.)

[278] Landé, A. Quantum theory from non-quantal postulates. In [39]. (MR 21 #6961.)

[279] Landsberg, P. T. An algebra of observables. Philosophical Magazine, 1947, 38, 757-773. (MR 10, p. 224.)

[280] Levi, I. Putnam's three truth values. Philosophical Studies, 1959, 10, 65-69.

[281] Łoś, J. Semantic representation of the probability of formulas in formalized theories. Studia Logica, 1963, 14, 183-196. (Reprinted in [41].)

[282] Ludwig, G. Attempt of an axiomatic foundation of quantum mechanics and more general theories, II. Communications in Mathematical Physics, 1967, 4, 331-348. (MR 37 #5003.)

[283] Ludwig, G. Attempt of an axiomatic foundation of quantum mechanics and more general theories, III. Communications in Mathematical Physics, 1968, 9, 1-12. (MR 31 #5004.)

[284] Ludwig, G. Versuch einer axiomatischen Grundlegung der Quantenmechanik und allgemeinerer physikalischer Theorien. Zeitschrift für Physik, 1964, 181, 233-260. (MR 31 #5473.)

[285] Mackey, G. W. Quantum mechanics and Hilbert space. American Mathematical Monthly, 1957, 64(8, part 2), 45-57. (MR 20 #2608.)

[286] Maczyński, M. J. Boolean properties of observables in axiomatic quantum mechanics. Reports on Mathematical Physics, 1971, 2, 135-150. (MR 44 #5012.)

[287] Maczyński, M. J. Functional properties of quantum logics. International Journal of Theoretical Physics, 1974, 11, 149-156.

[288] Maczyński, M. J. Orthogonality postulate in axiomatic quantum mechanics. International Journal of Theoretical Physics, 1973, 8, 353-360.

[289] Maczyński, M. J. Quantum families of Boolean algebras. Bulletin de l'Académie Polonaise des Sciences, 1970, 18, 93-96. (MR 41 #9503.)

[290] Maczyński, M. J. A remark on Mackey's axiom system for quantum mechanics. Bulletin de l'Académie Polonaise des Sciences, 1967, 15, 583-587. (MR 36 #7387.)

[291] Maczyński, M. J., & Traczyk, T. Characterization of orthomodular partially ordered sets admitting a full set of states. Bulletin de l'Académie Polonaise des Sciences, 1972, 21, 3-8.

[292] Marlow, A. R. Implications of a new axiom set for quantum logic. In [22].

[293] McGrath, J. Only if quanta had logic. In P. D. Asquith & I. Hacking (Eds.), PSA 1978 (Vol. 1). East Lansing, Mich.: Philosophy of Science Association, 1976.

[294] McGrath, J. Review of Hooker's The logico-algebraic approach to quantum mechanics. Philosophy of Science, 1978, 45, 145-148.

[295] Mielnik, B. Quantum logic and evolution. Annales de l'Institut Henri Poincaré, 1968, 9, 1-5.

[296] Mielnik, B. Quantum logic: Is it necessarily orthocomplemented? In [34].

[297] Mielnik, B. Theory of filters. Communications in Mathematical Physics, 1969, 15, 1-46. (MR 41 #3004.)

[298] Mittelstaedt, P. On the interpretation of the lattice of subspaces of the Hilbert space as a propositional calculus. Zeitschrift für Naturforschung, 1972, 27a, 1358-1362.

[299] Mittelstaedt, P. Quantenlogik. Fortschritte der Physik, 1961, 9, 106-147. (MR 23 #B1381.)

[300] Mittelstaedt, P. Quantenlogische Interpretation orthokomplementärer quasimodularer Verbände. Zeitschrift für Naturforschung, 1970, 25a, 1773-1778. (MR 43 #4361.)

[301] Mittelstaedt, P. Quantum logic. In [23].

[302] Mittelstaedt, P. Über die Gültligkeit der Logik in der Natur. Naturwissenschaften, 1960, 47, 385-391.

[303] Mittelstaedt, P. Time dependent propositions and quantum logic. Journal of Philosophical Logic, 1977, 6, 463-472.

[304] Mittelstaedt, P. Untersuchungen zur Quantenlogik. Sitzungsberichte der Bayerisch Akademie der Wissenschaften (Mathematisch-naturwissenschaftliche Klasse), 1959, 321-386.

[305] Mittelstaedt, P., & Stachow, E. W. Operational foundation of quantum logic. Foundations of Physics, 1974, 4, 355-365.

[306] Mittelstaedt, P., & Stachow, E. W. The principle of excluded middle in quantum logic. Journal of Philosophical Logic, 1978, 7, 181-208.

[307] Moroz, B. Z. Formal systems that arise in the analysis of physical theories. Doklady Akademii Nauk SSSR, 1971, 198, 1018-1020. (MR 43 #8302.)

[308] Müller, H. Mehrwertige Logik und Quantenphysik. Physikalische Blätter, 1954, 10, 151-157.

[309] Nagel, E. Book review: Philosophical foundations of quantum mechanics by H. Reichenbach. Journal of Philosophy, 1945, 42, 437-444.

[310] Nagel, E. Professor Reichenbach on quantum mechanics: A rejoinder. Journal of Philosophy, 1946, 43, 247-250.

[311] Neumann, J. von. _Strict and probability quantum logics_. (Unpublished manuscript in Von Neumann Collection at Princeton University Library.) Reviewed by A. Taub in _Collected works of John Von Neumann_ (Vol. 4). New York: Macmillan, 1962.

[312] Nikodým, O. M. Studies of some items of the lattice theory in relation to the Hilbert-Hermite space. _Rendiconti del Seminario Matematico dell'Università di Padova_, 1969, _42_, 27-122. (_MR_ 41 #3003.)

[313] Ochs, W. On the covering law in quantal proposition systems. _Communications in Mathematical Physics_, 1972, _25_, 245-252. (_MR_ 45 #6280.)

[314] Ochs, W. On the foundation of quantal proposition systems. _Zeitschrift für Naturforschung_, 1972, _27a_, 893-900. (_MR_ 48 #13021.)

[315] Park, J. L., & Margenau, H. The logic of noncommutability of quantum-mechanical operators and its empirical consequences. In [73].

[316] Park, J. L., & Margenau, H. Simultaneous measurability in quantum theory. _International Journal of Theoretical Physics_, 1968, _1_, 211-283.

[317] Pauli, W. Reviewing study of Hans Reichenbach's _Philosophical foundations of quantum mechanics_. In [46].

[318] Piron, C. Axiomatique quantique. _Helvetica Physica Acta_, 1964, _37_, 439-468. (_MR_ 37 #3894.)

[319] Piron, C. Observables in general quantum theory. In [27].

[320] Piron, C. On the foundations of quantum physics. In [34].

[321] Piron, C. Survey of general quantum physics. _Foundations of Physics_, 1972, _2_, 287-314. (Reprinted in [42].)

[322] Pool, J. C. T. Baer*-semigroups and the logic of quantum mechanics. _Communications in Mathematical Physics_, 1948, _9_, 118-141. (Reprinted in [42].) (_MR_ 34 #5005.)

[323] Pool, J. C. T. _Events, observables, and operations and the mathematical approach to quantum theory_. Unpublished manuscript, Argonne National Laboratories, Argonne, Ill., no date.

[324] Pool, J. C. T. Semi-modularity and the logic of quantum mechanics. _Communications in Mathematical Physics_, 1968, _9_, 212-228. (Reprinted in [42].) (_MR_ 38 #1879.)

[325] Popper, K. R. Birkhoff and Von Neumann's interpretation of quantum mechanics. _Nature_, 1968, _219_, 682-685.

[326] Popper, K. R. Quantum theory, quantum logic and the calculus of probability. In [2].

[327] Prugovečki, E. An axiomatic approach to the formalism of quantum mechanics: I, II. _Journal of Mathematical Physics_, 1966, _7_, 1054-1069; 1070-1096. (_MR_ 36 #1151.)

[328] Prugovečki, E. A formalism for generalized quantum mechanics. _Journal of Mathematical Physics_, 1966, _7_, 1680-1696. (_MR_ 34 #1032.)

[329] Prugovečki, E. On a theory of measurement of incompatible observables in quantum mechanics. _Canadian Journal of Physics_, 1967, _45_, 2173-2219.

[330] Pulmanno, S. Superposition principle in quantum logics. _Communications in Mathematical Physics_, 1976, _59_, 47-51.

[331] Putnam, H. How to think quantum-logically. _Synthese_, 1974, _29_, 55-61. (Reprinted in [69].)

[332] Putnam, H. Is logic empirical? In [21].

[333] Putnam, H. Three-valued logic. *Philosophical Studies*, 1957, *8*, 73-80. (Reprinted in [42].)

[334] Pyatnitsym, V. N., & Meskov, V. S. On the status of logic in quantum mechanics. *Teorie a Metoda*, 1972, *4*, 111-129.

[335] Randall, C. H. *A mathematical foundation for empirical science--With special reference to quantum theory--Part 1, A calculus of experimental propositions*. Knolls Atomic Power Lab Report KAPL-3147, June 1966.

[336] Reichenbach, H. Les fondements logiques de la théorie des quanta: Utilisation d'une logique à trois valeurs. In P. Destouches-Février (Ed.), *Actes du 2ème Colloque International de Logique Mathématique*. Paris: Gauthier-Villars, 1952.

[337] Reichenbach, H. Reply to Ernest Nagel's criticism of my views on quantum mechanics. *Journal of Philosophy*, 1946, *43*, 239-247.

[338] Reiser, O. L. Non-Aristotelian logic and the crisis in science. *Scientia*, 1937, *61*, 137-150.

[339] Richter, E. Bemerkungen zur Quantenlogik. *Philosophia Naturalis*, 1964, *8*, 225-231.

[340] Rosser, J. B. On the many-valued logics. *American Journal of Physics*, 1941, *9*, 207-212.

[341] Routley, R. Ultralogic as universal? *Relevance Logic Newsletter*, May 1977, *2*, 138-175.

[342] Ruttiman, G. T. On the logical structure of quantum mechanics. *Foundations of Physics*, 1970, *1*, 173-182.

[343] Salmon, W. C. Verifiability and logic. In [33].

[344] Scheibe, E. Popper and quantum logic. British Journal for the Philosophy of Science, 1974, 25, 319-328.

[345] Scheurer, P. B. Logique fermionique et logique bosonique. International Logic Review, 1972, 3, 188-206. (MR 50 #9194.)

[346] Schröter, J. A note concerning propositions in quantum mechanics. Annals of Physics, 1970/1971, 25, 243-245. (MR 42 #8783.)

[347] She, C. Y., & Heffner, H. Simultaneous measurement of noncommuting observables. Physical Review, 1966, 152, 1103-1110.

[348] Sherman, S. On Segal's postulates for general quantum mechanics. Annals of Mathematics, 1956, 64, 593-601. (MR 18, p. 625.)

[349] Shimony, A. Filters with infinitely many components. Foundations of Physics, 1972, 2, 325-328.

[350] Smith, G. J. Review of Bub's Interpretation of quantum mechanics. International Philosophical Quarterly, 1978, 18.

[351] Sotirov, V. K. Grounds for quantum logic. Doklady Bulgarska Akademiia na Naukite, 1972, 25, 7.

[352] Specker, E. P. The logic of propositions which are not simultaneously decidable. Dialectica, 1960, 14, 239-246. (Reprinted in [42].)

[353] Stachel, J. The "logic" of "quantum logic." In [23].

[354] Stachow, E. W. Completeness of quantum logic. Journal of Philosophical Logic, 1976, 5, 237-280.

[355] Stenius, E. The semantic status of laws of logic. In [67].

[356] Strauss, M. Foundations of quantum mechanics. In [68]. (Reprinted in [42]; original German article in [48].)

[357] Strauss, M. The logic of complementarity and the foundation of quantum theory. In [41]. (Trans. from "Zur Begündung der statistischen Transformationstheorie der Quantenphysik." _Sitzungsberichte der Akademie der Wissenschaften, Berlin_, 1936, _27_, 90-113.)

[358] Strauss, M. Logics for quantum mechanics. _Foundations of Physics_, 1973, _3_, 265-276. (_MR_ 50 #3786.)

[359] Strauss, M. Mathematics as logical syntax--A method to formalize the language of a physical theory. In [42].

[360] Suppes, P. Logics appropriate to empirical theories. In [10]. (Reprinted in [42].) (_MR_ 34 #7133.)

[361] Suppes, P. The probabilistic argument for a nonclassical logic of quantum mechanics. _Philosophy of Science_, 1966, _33_, 14-21. (Reprinted in [42].) (_MR_ 35 #6415.)

[362] Suppes, P., & McKinsey, J. C. C. Review of Destouches-Février, _La structure des théories physiques_. _Journal of Symbolic Logic_, 1954, _19_, 52-55.

[363] Turner, J. Violation of the quantum ordering of propositions in hidden variable theories. _Journal of Mathematical Physics_, 1968, _9_, 1411-1415.

[364] Turquette, A. R. Review of Reichenbach's _Philosophical foundations of quantum mechanics_. _Philosophical Review_, 1945, _54_, 513-516.

[365] Umegaki, H. Conditional expectation in an operator algebra. Tôhoku Mathematical Journal, 1954, 6, 171-181; 1956, 8, 86-100.

[366] Varadarajan, V. S. Probability in physics and a theorem on simultaneous observability. Communications on Pure and Applied Mathematics, 1962, 15, 189-217. (Corrected in Communications on Pure and Applied Mathematics, 1965, 18, 757; reprinted in [41].)

[367] Waismann, F. Are there alternative logics? Proceedings of the Aristotelian Society, 1946, 46, 103ff.

[368] Watanabe, S. Algebra of observation. Progress of Theoretical Physics, 1966, 37 (Suppl.), 350-367.

[369] Weizsäcker, C. F. von. Classical and quantum descriptions. In [51].

[370] Weizsäcker, C. F. von. Komplementarität und Logik. Naturwissenschaften, 1955, 42, 548-555.

[371] Weizsäcker, C. F. von. Komplementarität und Logik II. Zeitschrift für Naturforschung, 1958, 13a, 245-253.

[372] Weizsäcker, C. F., Scheibe, E., & Süssmann, G. Komplementarität und Logik III. Zeitschrift für Naturforschung, 1958, 13a, 705-721.

[373] Wessels, L. Laws and meaning postulates in van Fraassen's view of theories. Paper presented at the 1974 Philosophy of Science Association Meeting, University of Notre Dame, Indiana, 1974.

[374] Wilbur, W. J. Characterizing standard quantum logics--Preliminary report. Notices of the American Mathematical Society, 1976, 23A, 14.

[375] Wilbur, W. J. Quantum logic and locally convex spaces. Transactions of the American Mathematical Society, 1975, 207, 343-360.

[376] Zawirski, Z. Attempts at application of many-valued logic to contemporary natural science. Sprawozdania Poznanskiego Towarzystwa, 1931, 2.

[377] Zeman, J. J. Generalized normal logic. Journal of Philosophical Logic, 1978, 7, 225.

[378] Zeman, J. J. Quantum logic with implication. Journal of Symbolic Logic, 1974, 39, 391.

[379] Zierler, N. Axioms for non-relativistic quantum mechanics. Pacific Journal of Mathematics, 1961, 11, 1151-1169. (Reprinted in [42].)

[380] Zierler, N. On lattice of closed subspaces of Hilbert space. Pacific Journal of Mathematics, 1966, 19, 583-586.

[381] Zierler, N., & Schlesinger, M. Boolean embeddings of orthomodular sets and quantum logic. Duke Mathematical Journal, 1965, 32, 251-262. (Reprinted in [42].) (MR 30 #5704.)

"MR 25 #5705" at the end of an article's or book's entry indicates that a short review of it can be found in Mathematical Reviews, volume 25, number 5705. Old volumes do not number the reviews, and a page number is indicated instead.

This bibliography includes the relevant materials from Don Nilson's "Bibliography on the History and Philosophy of Quantum Physics" in Logic and Probability in Quantum Mechanics, edited by P. Suppes. I am also indebted to Don Nilson for additional references to which he directed me.

NAME INDEX

SUBJECT INDEX